Applied Surface Mount Assembly

Applied Surface Mount Assembly

A Guide to Surface Mount
Materials and Processes

Robert Rowland
Paul Belangia

VNR VAN NOSTRAND REINHOLD
_____ New York

To our wives and parents

Copyright © 1993 by Van Nostrand Reinhold

Library of Congress Catalog Card Number 92-30337
ISBN 0-442-00727-2

All rights reserved. No part of this work covered by the copyright hereon may be reproduced or used in any form or by any means—graphic, electronic, or mechanical, including photocopying, recording, taping, or information storage and retrieval systems—without written permission of the publisher.

Manufactured in the United States of America.

Van Nostrand Reinhold
115 Fifth Avenue
New York, New York 10003

Chapman and Hall
2–6 Boundary Row
London, SE 1 8HN

Thomas Nelson Australia
102 Dodds Street
South Melbourne 3205
Victoria, Australia

Nelson Canada
1120 Birchmount Road
Scarborough, Ontario M1K 5G4, Canada

16 15 14 13 12 11 10 9 8 7 6 5 4 3 2 1

Library of Congress Cataloging-in-Publication Data

Rowland, Robert J., 1955—
   Applied surface mount assembly : a guide to surface mount materials and processes / Robert J. Rowland, Paul Belangia.
     p.   cm.
   Includes bibliographical references and index.
   ISBN 0-442-00727-2
   1. Printed circuits—Design and construction. 2. Surface mount technology.   I. Belangia, Paul, 1950—   . II. Title.
TK7868.P7R66 1992
621.3815′31—dc20   92-30337
                                                            CIP

Contents

Foreword xi

Acknowledgments xiii

Introduction xv

Chapter 1 **Introduction to Surface Mount Technology 1**

GLOSSARY 1
1.0 INTRODUCTION 2
1.1 WHAT IS SURFACE MOUNT TECHNOLOGY? 2
1.2 THE HISTORY OF SURFACE MOUNT TECHNOLOGY 3
1.3 WHY USE SURFACE MOUNT TECHNOLOGY? 4
1.4 CLASSIFICATIONS OF SURFACE MOUNT ASSEMBLIES 5
1.5 BASIC PROCESSES AND PROCESS FLOWS 7

Chapter 2 **Surface Mount Components and Component Packaging 9**

GLOSSARY 9
2.0 INTRODUCTION 11
2.1 CHIP COMPONENTS 11
2.2 ACTIVE COMPONENTS—PLASTIC 21

2.3 ACTIVE COMPONENTS—CERAMIC   39
2.4 MISCELLANEOUS COMPONENTS   40
2.5 COMPONENT PACKAGING   42

Chapter 3 Surface Mount Printed Circuit Boards   47

GLOSSARY   47
3.0 INTRODUCTION   49
3.1 MATERIAL REQUIREMENTS   49
3.2 PHYSICAL REQUIREMENTS   51
3.3 PLATING REQUIREMENTS   52
3.4 SOLDER MASK REQUIREMENTS   53
3.5 SILK SCREEN REQUIREMENTS   55
3.6 PRINTED CIRCUIT BOARD PANELS   55

Chapter 4 Designing for Assembly   59

GLOSSARY   59
4.0 INTRODUCTION   59
4.1 WHAT IS MANUFACTURABILITY ALL ABOUT?   60
4.2 COMPONENT SELECTION   60
4.3 COMPONENT PACKAGING   61
4.4 COMPONENT SPACING   61
4.5 COMPONENT LAYOUT   62
4.6 PAD AND HOLE SIZE   63
4.7 CONDUCTOR SIZE AND ROUTING   64
4.8 PCB AND ARRAY DESIGN   64

Chapter 5 Soldering Materials and Related Issues   67

GLOSSARY   67
5.0 INTRODUCTION   68
5.1 SURFACE MOUNT ADHESIVES   68
5.2 SOLDERING FLUX   69
5.3 SOLDER   72
5.4 SOLDER PASTE   75
5.5 SOLDERING BASICS   78
5.6 COMPONENT SURFACE FINISH   80
5.7 SUBSTRATE SURFACE FINISH   81

Contents   vii

Chapter 6   **Adhesive and Solder Paste Application Methods   83**

GLOSSARY   83
6.0 INTRODUCTION   83
6.1 PRINTING   84
6.2 SCREEN AND STENCIL FABRICATION   85
6.3 SQUEEGEE BLADES   89
6.4 PRINTING SYSTEMS   92
6.5 PRINTING ADHESIVES   98
6.6 PRINTING SOLDER PASTE   100
6.7 PROCESS PARAMETERS   102
6.8 PRINTED CIRCUIT BOARD DESIGN   103
6.9 PRINTING DEFECTS   104
6.10 DISPENSING   106
6.11 DISPENSING METHODS AND EQUIPMENT   106
6.12 DISPENSING DEFECTS   110
6.13 VISION ALIGNMENT   112

Chapter 7   **Component Placement   115**

GLOSSARY   115
7.0 INTRODUCTION   115
7.1 PLACEMENT EQUIPMENT CLASSIFICATION   117
7.2 PLACEMENT HEADS   124
7.3 POSITIONING SYSTEMS   125
7.4 COMPONENT CENTERING   125
7.5 COMPONENT FEEDERS   128
7.6 COMPONENT TAPING   131

Chapter 8   **Reflow Soldering and Adhesive Curing   135**

GLOSSARY   135
8.0 INTRODUCTION   136
8.1 PROFILING ISSUES   137
8.2 VAPOR-PHASE REFLOW   141
8.3 INFRARED REFLOW: LAMP IR AND PANEL IR   146
8.4 FORCED AIR CONVECTION REFLOW   152
8.5 CONTROL SYSTEMS   156
8.6 IN-LINE TRANSFER SYSTEMS   157

viii　Contents

　　　　　　　　　　8.7 CONTROLLED ATMOSPHERES　159
　　　　　　　　　　8.8 TIME/TEMPERATURE PROFILING　159
　　　　　　　　　　8.9 ADHESIVE CURING　162

Chapter 9　　**Wave Soldering**　165

　　　　　　　　　　GLOSSARY　165
　　　　　　　　　　9.0 INTRODUCTION　166
　　　　　　　　　　9.1 TIME/TEMPERATURE PROFILES　166
　　　　　　　　　　9.2 COMPONENT LAYOUT　170
　　　　　　　　　　9.3 FLUX APPLICATION　172
　　　　　　　　　　9.4 PREHEAT　174
　　　　　　　　　　9.5 SOLDER APPLICATION　176
　　　　　　　　　　9.6 PCB HANDLING　181

Chapter 10　　**Cleaning**　185

　　　　　　　　　　GLOSSARY　185
　　　　　　　　　　10.0 INTRODUCTION　186
　　　　　　　　　　10.1 CONTAMINANTS　186
　　　　　　　　　　10.2 SOLVENT CLEANING　188
　　　　　　　　　　10.3 AQUEOUS CLEANING　191
　　　　　　　　　　10.4 SEMIAQUEOUS CLEANING　194
　　　　　　　　　　10.5 WASTE WATER MANAGEMENT　197
　　　　　　　　　　10.6 CLEANLINESS TESTING　197

Chapter 11　　**Rework**　201

　　　　　　　　　　GLOSSARY　201
　　　　　　　　　　11.0 INTRODUCTION　201
　　　　　　　　　　11.1 COMPONENT CONCERNS　202
　　　　　　　　　　11.2 CONTACT SOLDERING　203
　　　　　　　　　　11.3 HOT GAS SOLDERING　206
　　　　　　　　　　11.4 INFRARED SOLDERING　208
　　　　　　　　　　11.5 GENERAL REQUIREMENTS　210
　　　　　　　　　　11.6 WORKMANSHIP STANDARDS　211

Chapter 12　　**Manufacturing Operations**　213

　　　　　　　　　　GLOSSARY　213
　　　　　　　　　　12.0 INTRODUCTION　213
　　　　　　　　　　12.1 LOT-SIZE CONSIDERATIONS　214
　　　　　　　　　　12.2 SET-UP CONSIDERATIONS　217

12.3 PROCESS FLOW 218
12.4 AUTOMATED MATERIAL HANDLING 222
12.5 EQUIPMENT MAINTENANCE 226

Appendix A **Documentation** **229**

ORGANIZATIONS 229

Index 237

Foreword

Surface mount technology is evolving into the dominant electronic assembly method throughout the international electronics community. Its growth to dominance is only the result of hard work by dedicated champions who preserver to understand, establish, and expand the capabilities of a process intensive manufacturing technology such as surface mount technology. The authors of this text are such champions. They have spent many years implementing and improving the technology for their companies.

I believe this text will provide value to every engineer and manager involved with electronics assembly. It will provide confidence to those just beginning to implement SMT, and it will provide insight and solutions to the more experienced.

The authors provide the reader with a through explanation, from a user's perspective, of components, printed circuit boards, and the various assembly processes.

All in all, the text provides a much needed quality-oriented treatise of this exciting and complex topic.

Phil P. Marcoux
PPM Associates, Sunnyvale, CA

Acknowledgments

Many people have supported our efforts to write this book, for which we thank them deeply. First among them are our wives, Patt and Denise. We are indebted to Steve Chapman at Van Nostrand Reinhold and Kate Scully at Northeastern Graphic Services for helping us start and complete this project. We also want to thank the management at Teradyne and Compaq for supporting our desire to write this book.

The following individuals generously supplied information and/or carefully reviewed various portions of our text:

Jim Blankenhorn—SMT Plus
Greg Blount—Praegitzer
Scott Buttars—Compaq
Dave Eck—Solar Products
Curt Eppley—AMP
Chuck Germer—Teradyne
John Gin—Teradyne
Dave Heller—Heller Industries
Arlene Kormis—Pace
Dennis Lafreniere—Teradyne
Betty Martin—American Research
Phil Marcoux—PPM Associates
Robert Mathein—AVX
Laurie Morrow—Teradyne
Dave Murrin—Technical Devices
Mark Peo—Heller Industries
Mark Schwarz—SCM Metal Products
Randy Speelman—Heller Industries
John Stabley—Ismeca
R. J. Thompson—Kemet
Chunglim Yeo—Teradyne

Introduction

KNOWLEDGE IS WEALTH

Surface mount technology (SMT) continues to gain acceptance and market share in the electronics industry. In many products it is the dominant technology. As the scope of SMT increases, it becomes more and more difficult to keep abreast of the material and process changes. This book will help those individuals involved with the manufacture of surface mount assemblies to understand the various material and process issues.

This book is basically divided into two sections. The first four chapters, written by Paul, explain surface mount material and design issues. The following eight chapters, written by Robert, explain the various surface mount manufacturing processes.

One of the most important aspects of SMT is the proper use of terms, definitions, and nomenclature. We have taken a different approach to this topic by including a detailed glossary of appropriate terms and definitions at the beginning of each chapter. This allows the reader to reference the desired definition easily. To obtain the most benefit from each chapter, we recommend that the glossary be read carefully first.

Each chapter explains the basic issues as well as the more advanced subject matter. Without a complete understanding of SMT basics it is very difficult to understand the more advanced subjects. We believe it is important to learn how to walk before attempting to run. We have supplied the reader with as much information as possible relating to a particular topic. Our approach is to provide the reader with all of the possible material, equipment, and process alternatives. This allows the reader to select the best solution for his or her operation.

This book will be helpful to engineers, technicians, designers, and managers involved with the design and manufacture of surface mount assemblies. The main focus of this book is consumer electronics, primarily because this is our area of expertise. In one form or another, we have worked with every material and process covered in this book.

This book was written with two ideas in mind: first, to share our knowledge of surface mount technology with others; and second, to improve our understanding of surface mount technology. We have done our best to accomplish both.

1
Introduction to Surface Mount Technology

GLOSSARY

Class A A classification of a surface mount assembly that only uses through-hole components. It can be associated with Type 1 or Type 2 SMT.

Class B A classification of a surface mount assembly that only uses surface mount components. It can be associated with Type 1 or Type 2 SMT.

Class C A classification of a surface mount assembly that uses a mixture of through-hole and surface mount components. It can be associated with Type 1 or Type 2 SMT.

Dual In-Line Package (DIP) The dual in-line package is used in through-hole technology. It consists of two rows of leads extending at right angles.

Integrated Circuit (IC) A type of component that incorporates the integration of many transistors, capacitors, resistors, and other component types on a single silicon chip.

Land The metallic surface to which a surface mount lead is soldered.

Pad The metallic ring that surrounds the through-hole or via on an SMT board.

PCA Printed Circuit Assembly.

PCB Printed Circuit Board.

SMC Surface Mount Component.

SMT Surface Mount Technology.

Statistical Process Control (SPC) A technique used for the control of a process using statistics as a basis.

Type 1 A classification of SMT in which SMCs are only placed on one side of the PCB.

Type 2 A classification of SMT in which SMCs are placed on both sides of the PCB.

1.0 INTRODUCTION

This chapter will review the history of surface mount technology and its advantages and disadvantages. It also covers the various process types and classifications. Note that all dimensions and temperatures throughout the book are in metric units with U.S. units in parentheses.

1.1 WHAT IS SURFACE MOUNT TECHNOLOGY?

Surface Mount Technology (SMT) is the system or process used to place Surface Mount Components (SMCs) on a Printed Circuit Board (PCB). SMCs are microminiature leaded or leadless components that are soldered directly to lands (pads) on the surface of a PCB.

Until the mid-1980s SMCs were used primarily in low-volume hybrid circuits, mostly due to a lack of automated production equipment that could manufacture large printed circuit assemblies (PCA). However, recent advances in technology have produced an assortment of automated production equipment that will handle a variety of component configurations, population densities, substrate sizes, and production volumes. By utilizing SMCs on these new automated production machines, PCAs can be manufactured for less overall cost than PCAs that use conventional through-hole technology. The cost savings will be realized in such areas as lower PCB cost (50-75% smaller), lower labor costs, greater system capacity per cubic volume due to

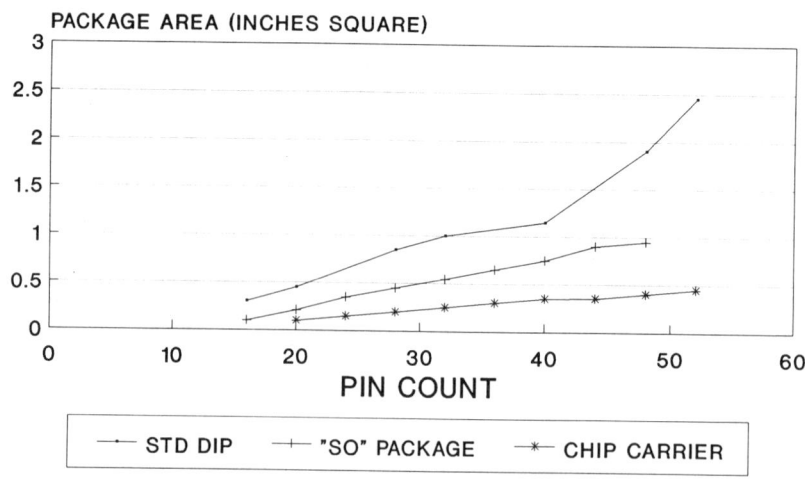

FIGURE 1.1. Comparison of SMT versus Through-Hole Technology.

smaller PCBs, reduced lead capacitance and inductance, more efficient manufacturing area, less rework, and faster circuitry, just to name a few.

Approximately 65% of the PCA cost is related to component size. As you can see by the data represented in Figure 1.1, it is easy to justify the use of surface mount technology. The return on investment (ROI) on the equipment can be as quick as two to six months!

1.2 THE HISTORY OF SURFACE MOUNT TECHNOLOGY

Surface mount technology has its roots in the 1950s. The early use of electronic components that did not use through-hole leads dates back to the use of flat packs for Hi-Rel military applications. In the 1960s more SMT components emerged to fit the needs of the limited hybrid market. Hybrid substrates are ceramic, necessitating that components be soldered on the surface of the substrate. In the 1970s the emerging Japanese electronics industry, being highly consumer oriented, was pushed to reduce costs. The types of products also required miniaturization to accommodate the market's needs. The first components to see wide usage were resistors and capacitors. They were used not only to reduce precious PCB space, but also to allow the use of high-speed placement equipment. The Japanese realized early on that handling a leadless cylindrical or rectangular component is much easier than forming, cutting, and clinching leads. In the late 1970s and early 1980s the

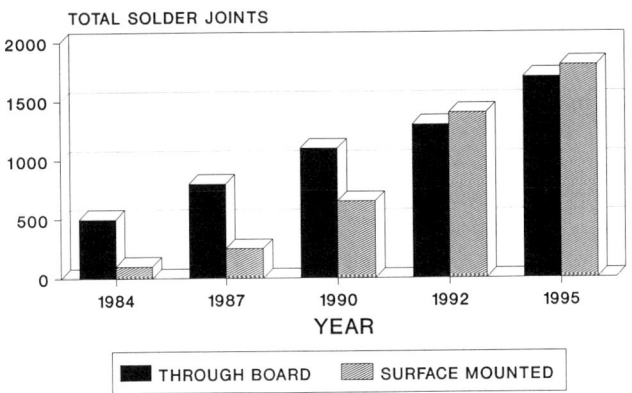

FIGURE 1.2. Growth Rate of SMT in the 1980s and 1990s. (Courtesy of Norplex Oak Inc.)

integrated circuit industry became very sophisticated and the circuitry became very complex. This dramatically increased the lead count, in many cases over 100. Using the dual in-line package (DIP) was now becoming a burden due to the large spaces required to accommodate these monsters.

Today the SMT industry is growing by leaps and bounds. SMT components are used in nearly all consumer and commercial products. The vast array of SMT components now covers many applications from power supplies to telecommunications. Figure 1.2 provides an example of the growth rate in the 80s and the projected growth rate in the 90s. The graph uses solder joints as a way of forecasting SMT growth.

1.3 WHY USE SURFACE MOUNT TECHNOLOGY?

1.3.1 Advantages of Surface Mount Technology

The product price/performance is vastly improved. This allows for a significant weight and size reduction. In addition to reaping the benefits of substantial weight and size reduction, SMT provides a cost effective solution to high pin count integrated circuits (ICs). In situations that require a very small package, surface mount technology is conducive to double-sided attachments to the PCB. An advantage that is sometimes not figured in is the cost advantage at the individual component level.

A higher product yield usually results since SMT manufacturing uses automation to place components. Automation improves quality through consistency. An automated set-up can also improve cycle time. The "inventory turns" can be thoroughly maximized. With the use of computerized manufacturing equipment, process control is much easier and more simplified. It allows for the use and implementation of statistical process control. Most SMT equipment presently available allows for high capacity even at the low price end.

Rework and repair of SMT boards is simplified. The standard tools are fine-tipped solder irons and/or hot air jets. Both tools require very small capital investment.

SMT components use up as little as 10% of the warehouse space required for through-hole components. Also, the placement equipment stores large volumes of components due to the advent of tape and reel packaging.

1.3.2 Disadvantages of Surface Mount Technology

There are a few minor disadvantages to SMT. The component commodity markets are slow to react to drastic changes such as this. Consequently all types of components are not available in surface mount packages. Board-level test is much more difficult. The spacing is much closer together, requiring smaller "pogo" test probes on the tester. The manufacturing of the surface mount PCB is much more complicated and requires very sophisticated placement equipment. This involves a very large capital outlay. One of the biggest disadvantages to a company just getting into SMT is the requirement to learn so many new processes.

1.4 CLASSIFICATIONS OF SURFACE MOUNT ASSEMBLIES

Surface mount assemblies are classified into two types: Type 1 and Type 2. They are then subclassified into three classes: Class A, Class B, and Class C. This was necessary to differentiate the location of the component mounting, one side or two sides, and the type of parts used in the assembly, be it through-hole and/or SMT. Class B and C also can be subclassified into simple and complex. This is documented in IPC-CM-770, "Printed Board Component Mounting." See Figure 1.3 for details.

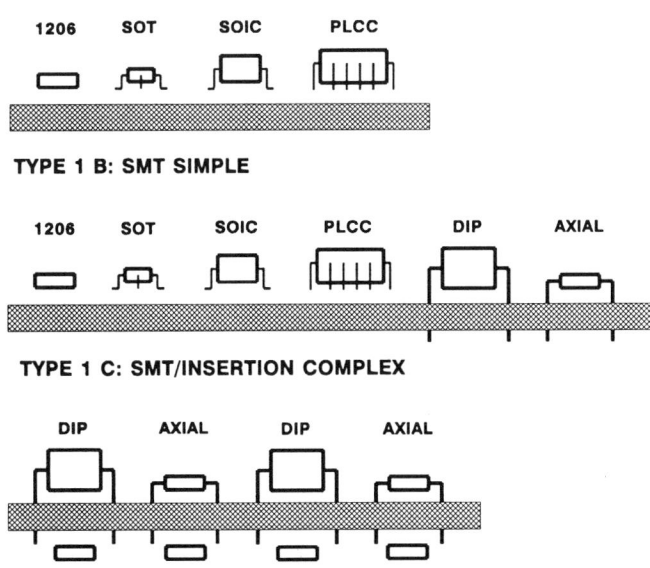

FIGURE 1.3a. Surface Mount Assembly Classification.

6   Applied Surface Mount Assembly

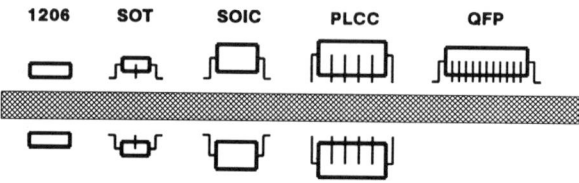

TYPE 2 B: SMT COMPLEX/FPT SIMPLE

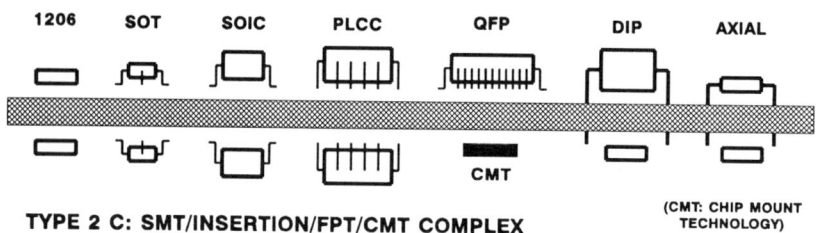

TYPE 2 C: SMT/INSERTION/FPT/CMT COMPLEX

(CMT: CHIP MOUNT TECHNOLOGY)

FIGURE 1.3b. Surface Mount Assembly Classification.

TYPE 1   Components mounted on only one side of the PCB
TYPE 2   Components mounted on both sides of the PCB

Class A   Through-hole component mounting only
Class B   Surface mount components only
Class C   A mixture of through-hole and surface mounting

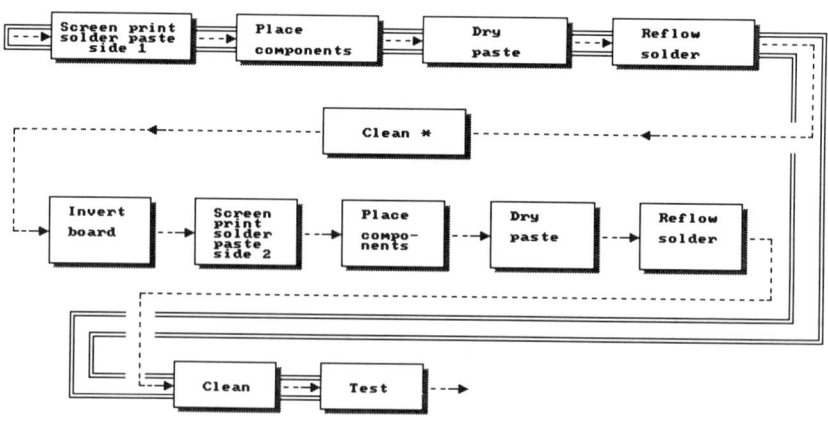

* Optional depending on final cleanliness specifications.

Type 1 SMT simple -(⇒) is process flow
Type 2 SMT complex - FPT simple - all blocks.

FIGURE 1.4.   Surface Mount Assembly Type 1B and 2B.

Introduction to Surface Mount Technology 7

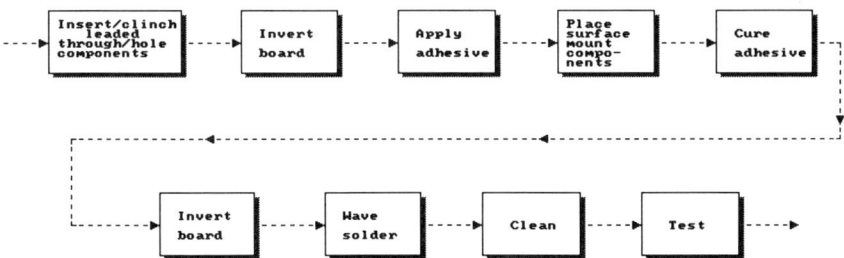

FIGURE 1.5. Surface Mount Assembly Type 2C Simple.

Thus, a Type 2B assembly is one that has only surface mount components on both sides of the board.

1.5 BASIC PROCESSES AND PROCESS FLOWS

The processes associated with the different SMT types differ by the assembly. Figure 1.4 shows an example of a Type 1B and 2B process flow.

An example of a Type 2C simple process flow is shown in Figure 1.5.

An example of a Type 2C complex process flow is shown in Figure 1.6.

There are several associations that develop standards for the American industry. The names and a list of the standards they develop can be found in Appendix A.

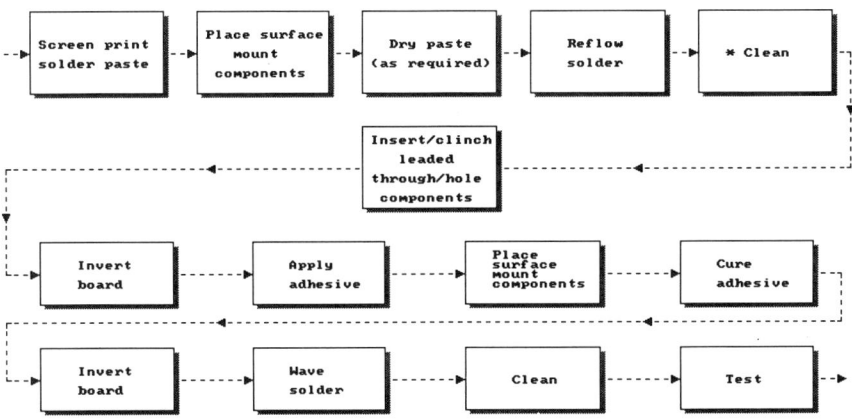

* Optional depending on final cleanliness specifications and solder flux used.

FIGURE 1.6. Surface Mount Assembly Type 2C Complex.

REFERENCES

1. Prasad, Ray. *Surface Mount Technology—Principles and Practices*. New York: Van Nostrand Reinhold, 1989.
2. Rowland, Robert. "Manufacturing Printed Circuit Assemblies that Utilize Surface Mounted Device Technology." Presented at the 21st Sperry Technical Symposium in Brainard, Minnesota, 1984.
3. Surface Mount Council, "Status of the Technology." IPC, Lincolnwood, IL, August 1992.
4. Tyler, Roger. "SMT: The Impact on Process and Market." *Circuits Assembly*, April 1991, pp. 74–80.
5. Yoo, Clarence. "New Concepts in SMT." *Circuits Assembly*, April 1991, pp. 60–64.

2
Surface Mount Components and Component Packaging

GLOSSARY

Bumpered Quad Flat Pack (BQFP)   An integrated circuit package with gull-wing leads on all four sides, with standard spacing between leads. Commonly the lead pitches are at 0.635mm (0.025″) centers, but lower pitches, for example, 0.508mm (0.020″), may also be used. The packages with lower pitches are generally referred to as fine-pitch packages. The bumpers denoted in the name are at each corner, the plastic extending out just past the length of the leads for protection. The quantity of leads ranges from 52 to 240.

Chip Component   Generic term for any two-terminal leadless surface mount passive device, such as a capacitor and resistor.

Ceramic Leaded Chip Carrier (CLCC)   A surface mount integrated circuit package similar to the leadless ceramic chip carrier, but with compliant leads attached to the package to prevent the problem associated with the coefficients of thermal expansion of the package and the substrate. The quantity of leads ranges from 16 to 124.

Coefficient of Thermal Expansion (CTE)   The fractional change in dimension of a material for a unit change in temperature.

Cylindrical Component   A package used for passive devices and diodes. It is cylindrical and comes in two main sizes: MLL34 and MLL41.

Dielectric   An insulating medium that occupies the region between two conductors.

Dual In-Line Package (DIP)   An integrated circuit package intended for through-hole mounting that has two rows of leads extending at right angles from the base, with standard spacing between leads and rows. Standard spacing is 2.54mm (0.100″).

Fine Pitch Surface mount packages with lead pitches equal to or less than 0.635mm (0.025").

Gull Wing Lead A lead configuration typically used on small outline packages where leads bend out. An end view of the package resembles a gull in flight.

J-Lead A lead configuration that has leads that bend underneath the body. It is typically used on plastic chip packages. A side view of the formed lead resembles the shape of the letter J.

Leadless Ceramic Chip Carrier (LCCC) A ceramic, hermetically sealed, integrated circuit package typically used for military applications. The package has metallized castellations on four sides for interconnecting to the substrate.

Lead Pitch The distance between successive centers of the leads of a component package. The lower the lead pitch, the smaller the package area for a given pin count in a package. The lead pitch in DIPs is 2.54mm (0.100"). In PLCCs it's 1.27mm (0.050").

Metal Electrode Leadless Face (MELF) A component package that is cylindrical and has metallization on its two ends. The package is commonly used for capacitors, resistors, and diodes.

Multilayer Chip Capacitor (MLC) The term used in the chip capacitor industry to describe surface mount ceramic chips.

Plastic Leaded Chip Carrier (PLCC) An integrated circuit package that has J-shaped leads on four sides, with 1.27mm (0.050") lead spacing. The quantity of leads ranges from 20 to 124.

Quad Pack Generic term for SMT packages with leads on all four sides. Most commonly used to describe packages with gull wing leads. Also known as a **quad flat pack (QFP)**.

Rectangular Component Generic term for any two-terminal leadless surface mount passive device, such as a capacitor and resistor. Also known as a chip component.

Shrink Small Outline Package (SSOP) An integrated circuit surface mount package similar to the SOIC with 8–30 leads. The body width is approximately 5.28mm (.208") and the pitch is 0.63mm (0.025").

Small Outline Integrated Circuit (SOIC) An integrated circuit surface mount package with two parallel rows of 8–16 gull-wing leads, with 1.27mm (0.050") spacing between leads and rows. The body width is approximately 3.81mm (0.150").

Small Outline J-Leaded (SOJ) An integrated circuit surface mount package with two parallel rows of 16–40 J leads, with 1.27mm (0.050") spacing between leads and rows. Typically used on memory devices.

Small Outline Large Integrated Circuit (SOLIC) An integrated circuit surface mount package with two parallel rows of 16–28 gull wing leads, with 1.27mm (0.050") spacing between leads and rows. The body width is approximately 7.63mm (0.300").

Small Outline Transistor (SOT) A discrete surface mount package that has two gull-wing leads on one side of the package and one on the other.

TapePak A type of packaging used to supply integrated circuits. The parts are supplied in a molded carrier ring.

Termination The metallization surfaces, or in some cases metal end clips, on the ends of passive chip components.

Thin Small Outline Package (TSOP) An integrated circuit surface mount package similar to the SOIC with 20–48 gull-wing leads. The width ranges from 6mm to 12mm (0.236" to 0.472"), including the leads. The pitch is 0.5mm (0.0197"). It is 1/2 of the regular SOIC thickness. Type I has leads extending from the narrow ends of the package. Type II has leads protruding from the wide sides of the package.

2.0 INTRODUCTION

The surface mount component industry has seen tremendous growth in the past ten years. It is becoming very difficult to keep up with the changes. One way to do this is to use industry standards, available from groups such as EIA, IPC, JEDEC, and EIAJ (See Appendix A). Unfortunately there are not standards for all components, and when there are standards they are not always followed. Japanese suppliers, for example, usually have their own designs, which do not conform to any standard. This is especially true for integrated circuits over 20 leads. The best advice is don't assume you know the component dimensions, especially for land pattern design, review the drawing, and make sure you know what the dimensions are.

This chapter reviews the common families of surface mount components and their land patterns and packaging. Reference IPC-SM-782 "Surface Mount Land Patterns" for more information.

2.1 CHIP COMPONENTS

Chip components encompass such components as resistors, capacitors, and inductors. They are the smallest surface mount packages. They have been more widely used in the automotive industries in the United States and Japan and in consumer products in Japan than other surface mount parts. The number indicates: length first (08 = 0.080") and width second (05 = 0.050").

2.1.1 Chip Resistors

Surface mount chip resistors come in three common sizes: 0805, 1206, and 1210. Inroads are starting to be made with 0603s and 0402s in high density applications, with the Japanese manufacturers leading the way. See Figure 2.1 for a size comparison.

Chip resistors are manufactured using an alumina substrate. The resistive element is then screened or sputtered on the substrate. The next process is to trim them to size. End terminations are then applied on three sides: top,

12 Applied Surface Mount Assembly

FIGURE 2.1. Chip Resistors Size Comparison. (Courtesy of KOA Speer Electronics, Inc.)

bottom, and end. The metallurgy of the terminations is silver paste, nickel, and tin, in that order. See Figure 2.2 for a cross section.

Value markings on surface mount resistors normally involve an extra cost since they are not usually marked. A three-digit marking system is used when markings are requested. See Figure 2.3 for an example of a resistor with the standard marking. The reels usually have the detail markings as required by the customer or per the standard set-up by the supplier. The value of the resistors should be checked as each new reel is installed on the placement system. This negates the need for individual part marking and the added cost. There are some companies that see the need to mark resistors for use in rework and/or service in the field.

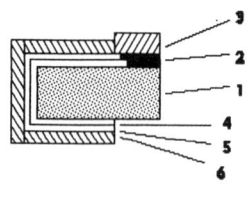

1: SUBSTATE CERAMIC (ALUMINA 96%)
2: RESISTIVE FILM RUTHENIUM OXIDE
3: PROTECTION FILM LEAD GLASS
4: INNER TERMINATION PALLADIUM/SILVER
5: INNER TERMINATION NICKEL
6: OUTER TERMINATION SOLDER

FIGURE 2.2. Chip Resistor Cross Section. (Courtesy of Rohm Co., LTD.)

MARKING:

Standard marking will be yellow characters on the element side. 5% tolerance parts will be marked with 3 characters, (2 significant characters and a multiplier)

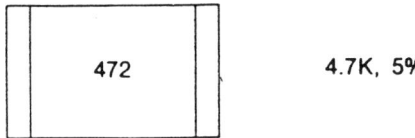

1% tolerance parts will be marked with 4 characters. (3 significant characters and a multiplier)

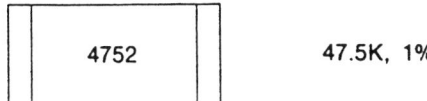

FIGURE 2.3. Resistor Standard Markings. (Courtesy of Rohm Co., LTD.)

Surface mount resistors can be procured as thin as 0.3mm (0.012″) for 0805s and 0603s. This could allow them to be placed in the center of a PGA socket as a space-saving effort.

See Table 2.1 for component dimensions. See Figure 2.4 for land patterns of surface mount chip resistors and Table 2.2 for the land pattern dimensions.

2.1.2 Chip Capacitors

Surface mount capacitors look similar to surface mount resistors except that capacitor terminations are on five sides: top, bottom, both sides, and the end.

TABLE 2.1 Chip Resistor Component Dimensions—Metric (Inches)

Size	Length	Width	Height
0805	1.90–2.10	1.14–1.35	.457–.660
	(.075–.083)	(.045–.053)	(.018–.026)
1206	3.00–3.20	1.45–1.70	.457–.660
	(.118–.126)	(.057–.067)	(.018–.026)
1210	3.00–3.40	2.39–2.79	.457–.660
	(.118–.134)	(.094–.110)	(.018–.026)

14   Applied Surface Mount Assembly

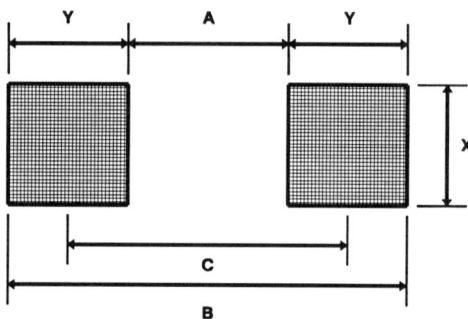

FIGURE 2.4.   Chip Resistor Land Patterns.

The metallurgy of the capacitor terminations is similar to the makeup of surface mount resistors. See Figure 2.5 for a cross-sectional view.

Monolithic surface mount capacitors are available in three different dielectric types: Z5U, X7R, and NPO or COG. Considering that the most common use of capacitors is for decoupling, X7R and Z5U dielectric are used because of their low cost. NPOs provide high stability for a wide range of temperatures, frequencies, and voltages, and of course cost substantially more.

Surface mount capacitors, also known as multilayer chip capacitors (MLCs), have quite a complex structure. A typical MLC will have 10 to 15 layers. Two different manufacturing methods are used. They are commonly referred to as the wet and dry processes. The dry process has been around for several years and was developed first. In that process, layers of green ceramic as thin as .025mm (0.001") are "layed up" one on top of the other, in a staggered fashion until the design parameters are achieved. One side of each layer has the screened electrode. Each individual part is then cut to size, baked, and fired. Terminations are then applied, which consist of palladium, nickel, and tin. See Figure 2.6 for a process flowchart of the dry process.

The wet process differs in that the layers are applied with wet green ceramic. This allows for much thinner layers to be applied. With the advent of the 0603 format the wet process is a necessity.

TABLE 2.2   Chip Resistor Land Pattern Dimensions—Metric (Inches)

Size	A	B	C	X	Y
0805	0.8 (.032)	3.8 (.150)	2.3 (.090)	1.4 (.055)	1.5 (.060)
1206	1.8 (.070)	5.0 (.200)	3.4 (.134)	1.6 (.063)	1.6 (.063)
1210	1.8 (.070)	5.0 (.200)	3.4 (.134)	2.6 (.102)	1.6 (.063)

Surface Mount Components and Component Packaging 15

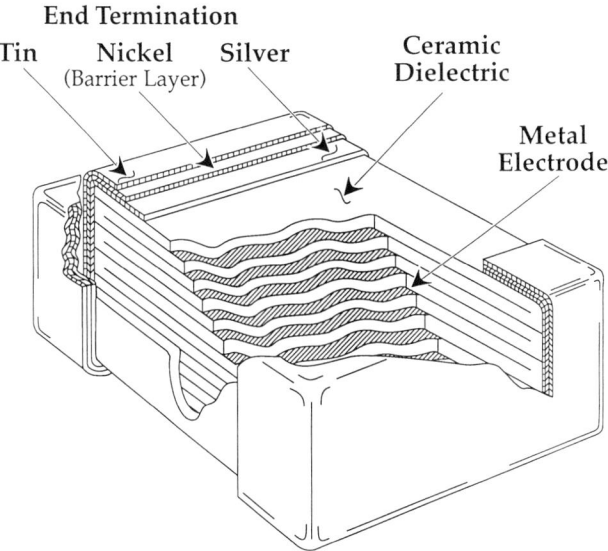

FIGURE 2.5. Chip Resistor Cross Sectional View. (Courtesy of Kemet Electronics Corp.)

See Table 2.3 for component dimensions. See Figure 2.7 for land patterns of surface mount chip capacitors and Table 2.4 for land pattern dimensions.

2.1.3 Molded Capacitors/Inductors

Molded capacitors are different from ceramic capacitors because they are usually polarized. Their value is also much higher than ceramics. The dielectric is tantalum. The capacitance values for tantalum capacitors varies from 0.1uF to 100uF and from 4 to 50 volts DC. They come in four different case sizes: A, B, C, and D. See Table 2.5 for component dimensions.

Since they are polarized they should never be used in a reversed polarity. They will explode causing a fire on the PCB. Ensure the polarity is correct when placing the reel on the placement system. See Figure 2.8 for a cross section of a molded tantalum capacitor.

Different manufacturers have various techniques of attaching the end terminations to the molded part. Some have a welded stub, while others have the exposed lead wrapped under the body of the capacitor. Some termination techniques are more reliable than others. The welded stub is being phased out due to problems associated with placement on the PCB.

See Figure 2.9 for land patterns of a molded capacitor and Table 2.6 for land pattern dimensions.

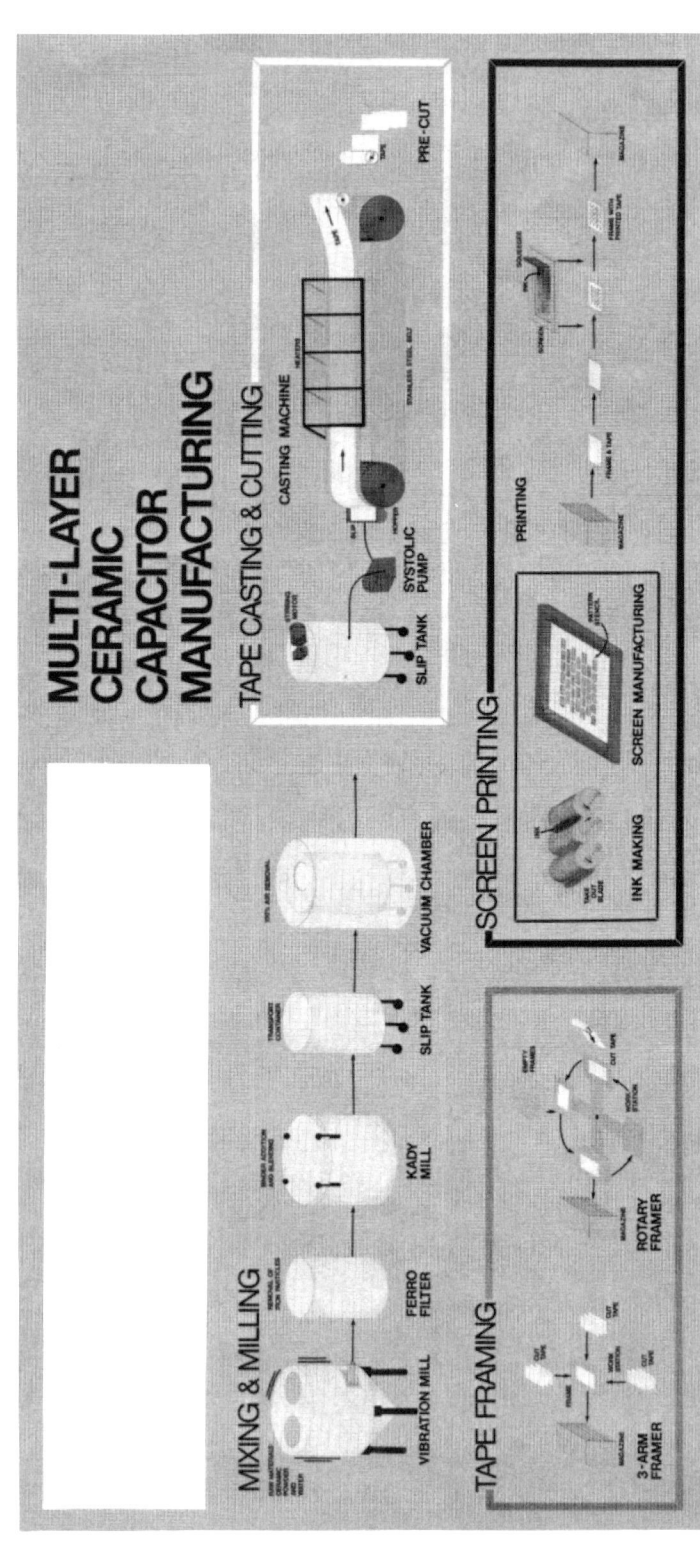

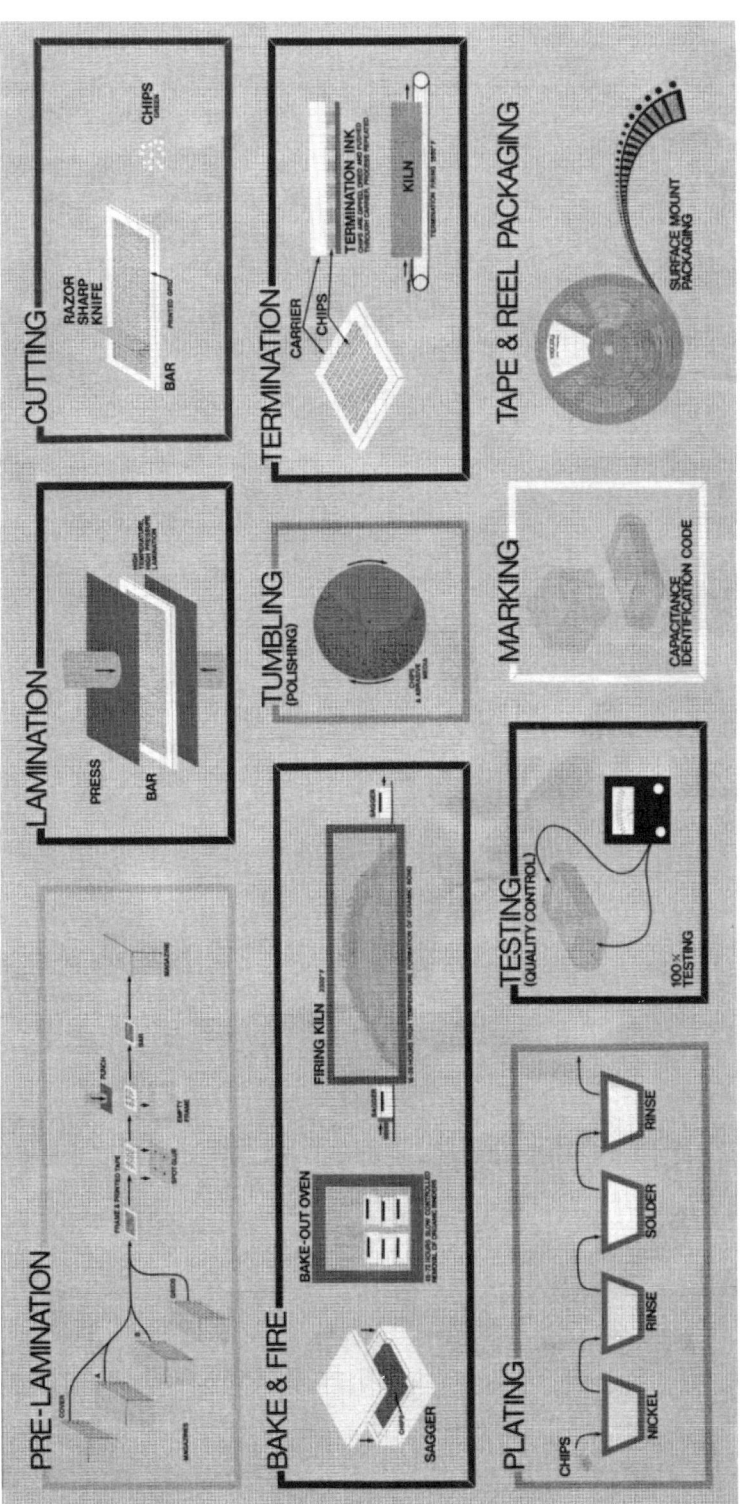

FIGURE 2.6. Chip Capacitor Dry Process Flowchart. (Courtesy of AVX/Kyocera Inc.)

TABLE 2.3 Chip Capacitor Component Dimensions—Inches

Size	Length	Width	Max Height
0603	.055–.071	.023–.039	.035
0805	.070–.087	.040–.055	.050
1206	.118–.134	.055–.070	.060
1210	.118–.134	.090–.106	.067
1812	.166–.190	.118–.134	.067
1825	.166–.190	.236–.268	.067

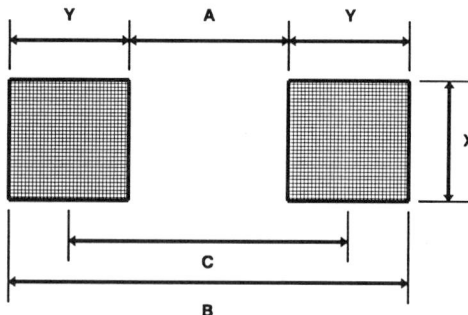

FIGURE 2.7. Chip Capacitor Land Patterns.

TABLE 2.4 Chip Capacitor Land Pattern Dimensions—Metric (Inches)

Size	A	B	C	X	Y
0603	1.0 (.040)	2.0 (.080)	1.5 (.060)	1.0 (.040)	0.5 (.020)
0805	0.8 (.032)	3.8 (.150)	2.3 (.090)	1.4 (.055)	1.5 (.060)
1206	1.8 (.070)	5.0 (.200)	3.4 (.134)	1.6 (.063)	1.6 (.063)
1210	1.8 (.070)	5.4 (.213)	3.6 (.142)	2.6 (.102)	1.8 (.070)
1812	3.2 (.126)	6.8 (.268)	5.0 (.200)	3.2 (.126)	1.8 (.070)
1825	3.2 (.126)	6.8 (.268)	5.0 (.200)	6.6 (.260)	1.8 (.070)

Surface Mount Components and Component Packaging   19

TABLE 2.5   Molded Capacitor Component Dimensions—Metric (Inches)

Size	A	B	C	D
Length	2.99–3.40	2.54–3.71	5.69–6.30	6.81–7.59
	(.118–.134)	(.130–.146)	(.224–.248)	(.268–.299)
Height	1.39–1.78	1.70–2.11	2.21–2.79	2.49–3.10
	(.055–.070)	(.067–.083)	(.087–.110)	(.098–.122)
Width	1.27–1.78	2.59–2.99	2.89–3.50	3.99–4.60
	(.050–.070)	(.102–.118)	(.114–.138)	(.157–.181)

Molded inductors are available from a few suppliers. TDK is an industry leader in the manufacture of surface mount inductors. They have a unique molded inductor without windings. This is accomplished by using alternating layers of ferrite paste and conductive silver paste. Most molded inductors use a coil wound on ferrite, which is then molded for ease of placement using the placement system.

2.1.4 Cylindrical Components (MELFs)

Cylindrical components are known in the industry as MELFs, which stands for metal electrode leadless face. This package style can be used to manufac-

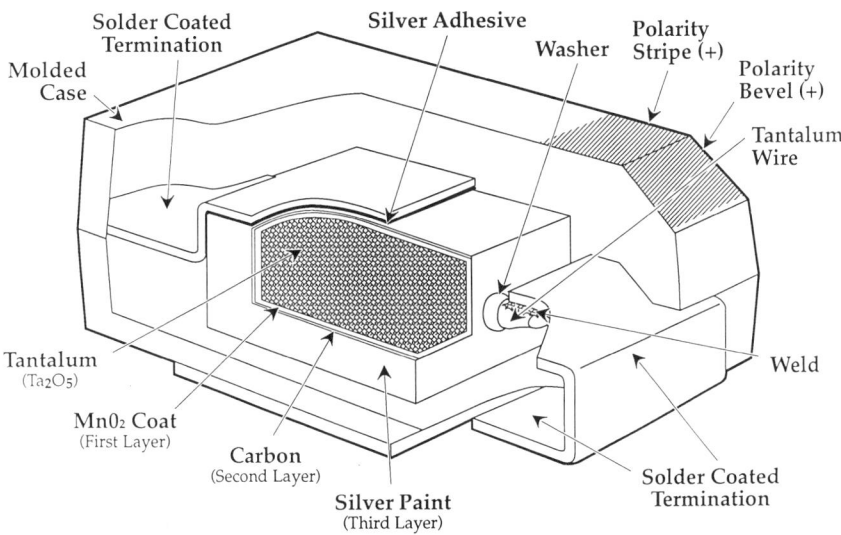

FIGURE 2.8.   Cross Section of a Molded Tantalum Capacitor. (Courtesy of Kemet Electronics Corp.)

20 Applied Surface Mount Assembly

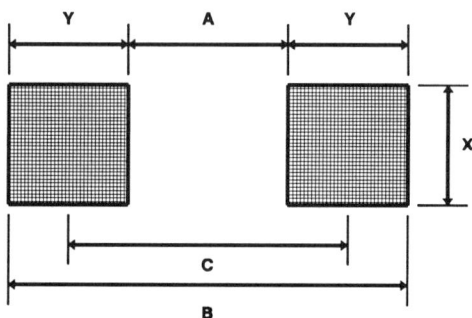

FIGURE 2.9. Molded Capacitor Land Patterns.

ture diodes, resistors, and ceramic and tantalum capacitors, but is most commonly used for resistors. They are basically the same raw resistor device used to make the leaded resistor, but without the leads and conformal coating. The two common sizes are MLL34 and MLL41. Since they are cylindrical, they have a tendency to roll off any surface to which they are not attached. For this reason the use of the cylindrical components is decreasing.

See Table 2.7 for MELF component dimensions. See Figure 2.10 for MELF land patterns and Table 2.8 for land patterns dimensions.

TABLE 2.6 Molded Capacitor Land Pattern Dimensions—Metric (Inches)

Size	A	B	C	X	Y
A	0.8 (.032)	4.8 (.190)	2.8 (.110)	1.4 (.055)	2.0 (.080)
B	1.2 (.047)	5.2 (.205)	3.2 (.126)	2.4 (.095)	2.0 (.080)
C	3.2 (.126)	8.0 (.315)	5.6 (.220)	2.4 (.095)	2.4 (.095)
D	4.2 (.166)	9.4 (.370)	7.0 (.276)	2.6 (.102)	2.4 (.095)

TABLE 2.7 MELF Component Dimensions—Metric (Inches)

Size	Diameter	Length
MLL34	1.6 (0.063)	3.5 (0.138)
MLL41	2.49 (0.098)	5.0 (0.197)

Surface Mount Components and Component Packaging 21

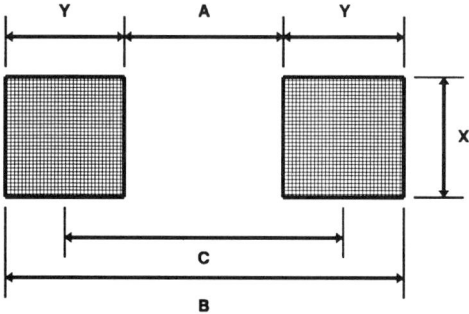

FIGURE 2.10. MELF Land Patterns.

TABLE 2.8 MELF Land Pattern Dimensions—Metric (Inches)

Size	A	B	C	X	Y
MLL34	2.2 (.087)	5.0 (.200)	3.6 (.142)	2.0 (.080)	1.4 (.055)
MLL41	3.4 (.134)	7.0 (.276)	5.2 (.205)	2.8 (.110)	1.8 (.070)

2.2 ACTIVE COMPONENTS—PLASTIC

The use of surface mount technology in the area of active components has allowed the manufacture of such products as Walkmans and half-height hard disk drives. These products would not be possible without the space savings offered by surface mount packages. See Figure 2.11 for a size comparison of leaded versus SMT.

2.2.1 Small Outline Transistors (SOT)

Small outline transistors (SOTs) are packaged in three surface mount styles: SOT23, SOT89, and SOT143. The SOT23 and the SOT89 have three leads, while the SOT143 has four leads. Diodes can also be assembled using these packages.

The SOT23 package is the most commonly used of the three packages. The package can be procured in three profiles: high, medium, or low. See Figure 2.12 for details.

If, for example, the low-profile part is called for, make sure the medium-profile or high-profile part is not substituted if the component is to be glued

22 Applied Surface Mount Assembly

	THROUGH-HOLE (TH)	SURFACE MOUNT (SMT)	FINE PITCH (FPT)	CHIP ON BOARD (COB)
PIN COUNT PER IC # of leads	8 to 64	8 to 244	38 to 804	Equal to IC pincount
DENSITY 16 Pin equivalent ICs per square inch	2 to 3	4 to 6	6 to 11	10 to 33

FIGURE 2.11. Surface Mount IC Packaging Size Comparisons.

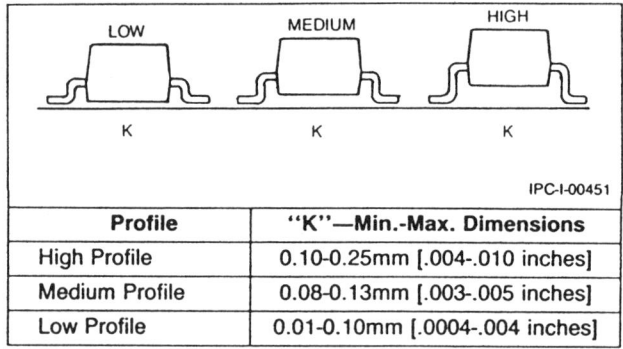

Profile	"K"—Min.-Max. Dimensions
High Profile	0.10-0.25mm [.004-.010 inches]
Medium Profile	0.08-0.13mm [.003-.005 inches]
Low Profile	0.01-0.10mm [.0004-.004 inches]

FIGURE 2.12. SOT 23 Package Size Comparison.

down. The glue dot might not be sufficient (tall) enough to touch and adhere to the taller part. The higher profile is typically used in applications that require thorough cleaning.

See Table 2.9 for the component dimensions. See Figure 2.13 for land patterns and Table 2.10 for land pattern dimensions.

The SOT89 package was designed to handle high-power transistors that the SOT23 could not. It has limited use in the industry.

TABLE 2.9 SOT23 Component Dimensions—Metric (Inches)

Dimension	MIN–MAX
"A" Length	2.80–3.05 (0.110–0.120)
"B" Width (T-T)	2.10–2.50 (0.083–0.098)
"C" Height	0.85–1.20 (0.033–0.048)
"D" Width (body)	1.20–1.40 (0.048–0.055)

Surface Mount Components and Component Packaging 23

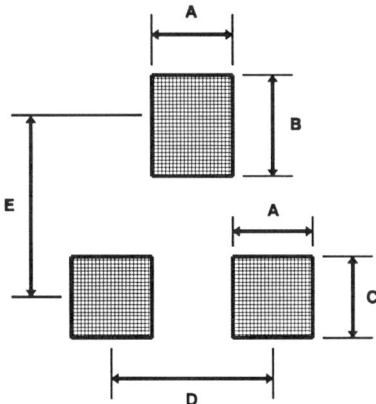

FIGURE 2.13. SOT23 Land Patterns.

See Table 2.11 for the component dimensions. See Figure 2.14 for the land patterns and Table 2.12 for land pattern dimensions.

The SOT143 looks similar to a SOT23, except it has 4 leads. They are starting to become more popular in the miniature radio frequency (RF) applications. Also, cleaning is not a problem for this package.

See Table 2.13 for the component dimensions. See Figure 2.15 for the land patterns and Table 2.14 for land pattern dimensions.

TABLE 2.10 Land Pattern Dimensions—Metric (Inches)

Size	A	B	C	D	E
SOT23	1.0 (.040)	1.4 (.055)	1.2 (.048)	2.0 (.080)	2.2 (.088)

TABLE 2.11 SOT89 Component Dimensions—Metric (Inches)

Dimension	MIN–MAX
"A" Length	4.40–4.60 (0.174–0.182)
"B" Width (T-T)	3.94–4.25 (0.156–0.167)
"C" Height	1.40–1.60 (0.056–0.062)
"D" Width (body)	2.29–2.60 (0.090–0.102)

24   Applied Surface Mount Assembly

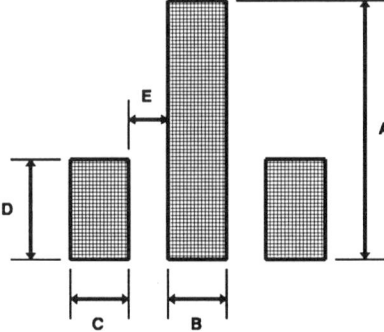

FIGURE 2.14. SOT89 Land Patterns.

The DPAK is a higher wattage version of the SOT package. It can handle what is equivalent to the TO-252 package in the leaded package format.

See Table 2.15 for the component dimensions. See Figure 2.16 for the land patterns and Table 2.16 for land pattern dimensions.

2.2.2 Small Outline (SO) Packages with Gull Wing Leads

The small outline integrated circuit (SOIC) is a small outline package that is similar to the DIP, but is exactly half the size; the body width is 3.81mm

TABLE 2.12   SOT89 Land Pattern Dimensions—Metric (Inches)

Size	A	B	C	D	E
SOT89	5.4 (.213)	1.0 (.040)	1.0 (.040)	1.4 (.055)	0.5 (.020)

TABLE 2.13   SOT143 Component Dimensions—Metric (Inches)

Dimension	MIN–MAX
"A" Length	2.80–3.10 (0.110–0.122)
"B" Width (T-T)	2.10–2.60 (0.083–0.102)
"C" Height	0.85–1.20 (0.033–0.048)
"D" Width (body)	1.20–1.70 (0.048–0.067)

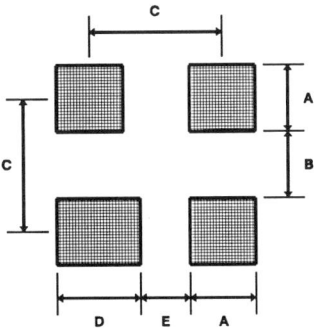

FIGURE 2.15. SOT143 Land Patterns.

(0.150″) instead of 7.62mm (0.300″) and the lead pitch is 1.27mm (0.050″) instead of 2.54mm (0.100″). There is a larger small outline package—the small outline large integrated circuit (SOLIC).

The two-sided, rectangular form of the SO package provides the best space utilization below 20 leads; above 20 leads four-sided packages such as the PLCC may provide better space utilization, depending on the conductor routing requirements.

Small outline packages come in two main body sizes: 3.81mm (0.150″) and 7.62mm (0.300″). The 3.81 mm (0.150″) size is called an SOIC and the 7.62mm

TABLE 2.14 SOT143 Land Pattern Dimensions—Metric (Inches)

Size	A	B	C	D	E
SOT143	1.0 (.040)	1.0 (.040)	2.0 (.080)	1.25 (.05)	0.75 (.03)

TABLE 2.15 DPAK Component Dimensions—Metric (Inches)

Dimension	MIN–MAX
"A" Length (T-T)	9–10 (.354–.393)
"B" Width (body)	6.3–6.7 (.248–.263)
"C" Height	2.2–2.5 (.087–.098)
"D" Lead Pitch	4.2–5.0 (.165–.197)

26 Applied Surface Mount Assembly

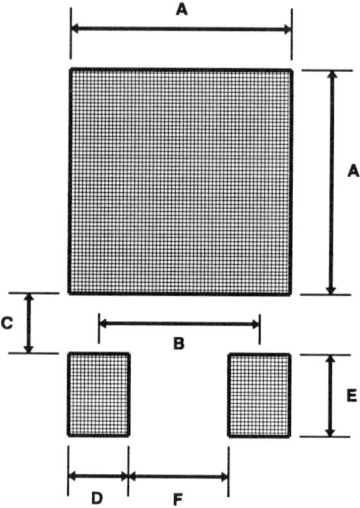

FIGURE 2.16. DPAK Land Patterns.

(0.300") size is called the SOLIC. The L means large. The 14- and 16-pin devices can be bought in either of the two package sizes.

See Table 2.17 for component dimensions. See Figure 2.17 for land patterns and Table 2.18 for land pattern dimensions.

2.2.3 Small Outline (SO) Packages with J Leads

The small outline J lead (SOJ) is typically used for packaging dynamic random access memory (DRAM) dies. The SOJ is similar to the PLCC package (see paragraph 2.2.4) in that it uses J leads instead of gull-wing leads. In the case of the SOJ the leads are only on two sides versus four sides for PLCCs. This is one of the package styles that has many variations due to the numerous Japanese suppliers of DRAMs. Body widths are

TABLE 2.16   DPAK Land Pattern Dimensions—Metric (Inches)

Size	A	B	C	D	E	F
DPAK	6.48 (.255)	4.40 (.173)	1.4 (.055)	1.4 (.055)	2.9 (.115)	3.0 (.120)

Surface Mount Components and Component Packaging   27

TABLE 2.17   SOIC Component Dimensions—Metric (Inches)

# Leads	Length	Width (T–T)	Height	Width (body)
8	4.75–5.0	5.8–6.2	1.3–1.75	3.8–4.0
	(.188–.197)	(.228–.244)	(.058–.069)	(.150–.158)
14	8.53–8.74	5.8–6.2	1.3–1.75	3.8–4.0
	(.366–.344)	(.228–.244)	(.058–.069)	(.150–.158)
16	9.78–10.0	5.8–6.2	1.3–1.75	3.8–4.0
	(.385–.395)	(.228–.244)	(.058–.069)	(.150–.158)
16L	10.11–10.41	10–10.6	2.4–2.6	7.4–7.6
	(.398–.413)	(.394–.419)	(.094–.102)	(2.91–2.99)
20L	12.6–13.0	10–10.6	2.4–2.6	7.4–7.6
	(.496–.512)	(.394–.419)	(.094–.102)	(2.91–2.99)
24L	15.21–15.6	10–10.6	2.4–2.6	7.4–7.6
	(.599–.614)	(.394–.419)	(.094–.102)	(2.91–2.99)
28L	17.7–18.11	10–10.6	2.4–2.6	7.4–7.6
	(.697–.713)	(.394–.419)	(.094–.102)	(2.91–2.99)

Note: The lead pitch is 1.27 (.050) in all cases.

7.62mm (0.300″), 8.9mm (0.350″) and 10.16mm (0.400″). The safest rule is to choose the supplier and then lay out the PCB with that particular supplier's land pattern suggestion.

See Table 2.19 for component dimensions. See Figure 2.18 for land patterns and Table 2.20 for land pattern dimensions.

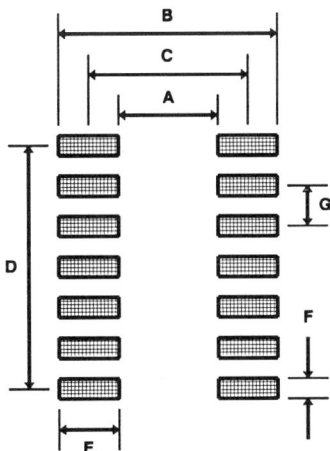

FIGURE 2.17.   SOIC Land Patterns.

TABLE 2.18 SOIC Land Pattern Dimensions—Metric (Inches)

# Leads	A	B	C	D	E	F
8	3.6 (.140)	7.6 (.300)	5.6 (.220)	3.8 (.150)	2.0 (.080)	.63 (.025)
14	3.6 (.140)	7.6 (.300)	5.6 (.220)	7.6 (.300)	2.0 (.080)	.63 (.025)
16	3.6 (.140)	7.6 (.300)	5.6 (.220)	8.9 (.350)	2.0 (.080)	.63 (.025)
16L	7.6 (.300)	11.6 (.46)	9.6 (.380)	8.9 (.350)	2.0 (.080)	.63 (.025)
20L	7.6 (.300)	11.6 (.46)	9.6 (.380)	11.4 (.45)	2.0 (.080)	.63 (.025)
24L	7.6 (.300)	11.6 (.46)	9.6 (.380)	14 (.550)	2.0 (.080)	.63 (.025)
28L	7.6 (.300)	11.6 (.46)	9.6 (.380)	16.5 (.65)	2.0 (.080)	.63 (.025)

Note: The lead pitch (G) is 1.27 (.050) in all cases.

2.2.4 Plastic Leaded Chip Carrier (PLCC)

The plastic leaded chip carrier (PLCC) was developed from the ceramic leaded chip carrier because the market required a lower cost package for consumer applications. The PLCC was the logical solution. It has leads bent under in the form of a J. The leads are on a 1.27mm (0.050″) pitch. An equal number of leads are on all four sides in a square format, except for the 18 and 32 position sizes, which are rectangular.

The pin one numbering system is different from SOICs in that the number

TABLE 2.19 SOJ Component Dimensions—Metric (Inches)

# Leads	Length
14	9.4–9.65 (.370–.380)
16	10.67–10.9 (.420–.430)
18	11.94–12.19 (.470–.480)
20	13.20–13.46 (.520–.530)
24	15.75–16.00 (.620–.630)
28	18.29–18.54 (.720–.730)
32	20.83–21.08 (.820–.830)
40	25.91–26.16 (1.02–1.03)

Note: The lead pitch (G) is 1.27 (.050) in all cases.

Surface Mount Components and Component Packaging 29

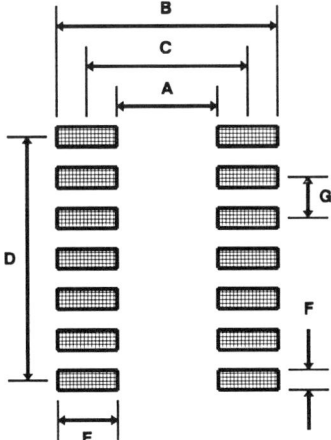

FIGURE 2.18. SOJ Land Patterns.

one mark is in the center of one side. The integrated circuit manufacturer usually molds the pin number mark in the plastic in a conspicuous manner. Unfortunately there are a few manufacturers that use the bevel in the corner as pin number one. Do not assume which pin is number one, ask the manufacturer.

TABLE 2.20 SOJ Land Pattern Dimensions—Metric (Inches)

# Leads	A	B	C	D	E	F
14	4.9	8.9	6.9	7.6	2.0	.63
	(.190)	(.350)	(.270)	(.300)	(.080)	(.025)
16	4.9	8.9	6.9	8.9	2.0	.63
	(.190)	(.350)	(.270)	(.350)	(.080)	(.025)
18	4.9	8.9	6.9	10.2	2.0	.63
	(.190)	(.350)	(.270)	(.40)	(.080)	(.025)
20	4.9	8.9	6.9	11.4	2.0	.63
	(.190)	(.350)	(.270)	(.45)	(.080)	(.025)
24	4.9	8.9	6.9	14	2.0	.63
	(.190)	(.350)	(.270)	(.550)	(.080)	(.025)
28	4.9	8.9	6.9	16.5	2.0	.63
	(.190)	(.350)	(.270)	(.65)	(.080)	(.025)
32	10.9	14.9	12.9	19	2.0	.63
	(.43)	(.586)	(.51)	(.750)	(.080)	(.025)
40	10.9	14.9	12.9	24.1	2.0	.63
	(.43)	(.586)	(.51)	(.95)	(.080)	(.025)

Note: The lead patch (G) is 1.27 (.050) in all cases.

TABLE 2.21 PLCC Component Dimensions—Metric (Inches)

# Leads	Length	Width
20	9.78–10.03 (.385–.395)	9.78–10.03 (.385–.395)
22 Rec.	13.21–13.59 (.520–.535)	8.13–8.51 (.320–.335)
28 Sq.	12.32–12.57 (.485–.495)	12.32–12.57 (.485–.495)
28 Rec.	14.86–15.11 (.585–.595)	9.78–10.03 (.385–.395)
32 Rec.	14.86–15.11 (.585–.595)	12.32–12.57 (.485–.495)
44	17.40–17.65 (.685–.695)	17.40–17.65 (.685–.695)
52	19.94–20.19 (.785–.795)	19.94–20.19 (.785–.795)
68	25.02–25.27 (.985–.995)	25.02–25.27 (.985–.995)
84	30.10–30.35 (1.185–1.195)	30.10–30.35 (1.185–1.195)
100	35.18–35.43 (1.385–1.395)	35.18–35.43 (1.385–1.395)
124	42.80–43.05 (1.685–1.695)	42.80–43.05 (1.685–1.695)

Note: The lead pitch is 1.27 (.050) in all cases.

Larger PLCCs (>68 leads) should be baked to remove moisture and then dry-packed until assembly. See IPC-SM-786, "Testing and Handling of Surface Mount Plastic Packages Susceptible To Moisture-Induced Cracking," for details on cracks and the use of baking to remove moisture.

See Table 2.21 for component dimensions. See Figure 2.19 for land patterns and Table 2.22 for land pattern dimensions.

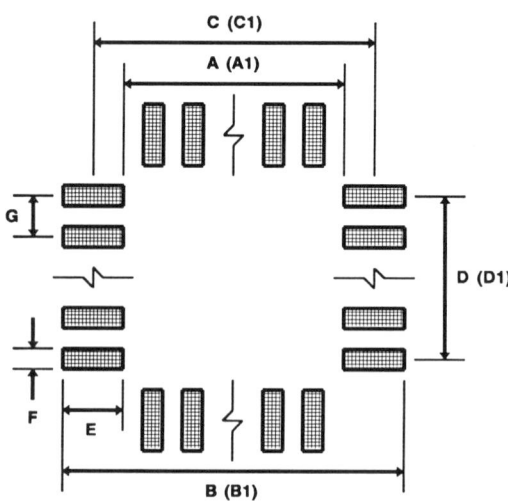

FIGURE 2.19. PLCC Land Patterns.

Surface Mount Components and Component Packaging   31

TABLE 2.22   PLCC Land Pattern Dimensions Metric (Inches)

# Leads	A/A1	B/B1	C/C1	D/D1	E	F
20	6.8 (.265)	10.8 (.425)	8.8 (.345)	5.08 (.20)	2.0 (.080)	.66 (.026)
22 rec.	5.4 (.210)	9.4 (.370)	7.4 (.290)	3.8 (.150)	2.0 (.080)	.66 (.026)
	10.6 (.415)	14.6 (.575)	12.6 (.495)	7.6 (.300)		
28 sq.	9.3 (.365)	13 (.525)	11.3 (.445)	7.6 (.300)	2.0 (.080)	.66 (.026)
28 rec.	6.8 (.265)	10.8 (.425)	8.8 (.345)	5.08 (.20)	2.0 (.080)	.66 (.026)
	12(.470)	16(.630)	14(.550)	10(.40)		
32 rec.	9.4 (.370)	13.4 (.530)	11.4 (.450)	7.6 (.300)	2.0 (.080)	.66 (.026)
	12(.470)	16(.630)	14(.550)	10.2 (.40)		
44	14.4 (.565)	18.4 (.725)	16.4 (.645)	12.7 (.50)	2.0 (.080)	.66 (.026)
52	17(.665)	21(.825)	19(.745)	15.2 (.60)	2.0 (.080)	.66 (.026)
68	22(.865)	26 (1.025)	24 (.945)	20.3 (.80)	2.0 (.080)	.66 (.026)
84	27.1 (1.065)	31.1 (1.225)	29.1 (1.145)	25.4 (1.000)	2.0 (.080)	.66 (.026)
100	32.2 (1.265)	36.2 (1.425)	34.2 (1.345)	30.5 (1.200)	2.0 (.080)	.66 (.026)
124	39.8 (1.565)	43.8 (1.725)	41.8 (1.645)	38.1 (1.500)	2.0 (.080)	.66 (.026)

Note: Position of dimensions A1–D1 is rotated 90 degrees on rectangular PLCCs.

2.2.5 Fine-Pitch Packages (QFP, BQFP, TSOP, SSOP, TapePak)

The quad flat pack (QFP) was developed in the late 1980s as a solution for integrated circuits that were pushing the limits of the PLCC package. Currently, the common lead pitch is 0.63mm (0.025"). The 0.508mm (0.020") lead pitch has been introduced while lead pitches below 0.508mm (0.020") are pending. The bumpered quad flat pack (BQFP) was introduced by JEDEC. JEDEC is the Joint Electronic Devices Engineering Council, which controls the standard. The bumpers allow the package to be supplied in tubes as well as matrix trays. The bumper also makes it easier to use the package in tape and reel. The standard package sizes in the United States are 52, 68, 84, 100, 132, and 196 leads.

TABLE 2.23 BQFP (JEDEC) Component Dimensions—Metric (Inches)

# Leads	Length	Width A	Width B
84	20.3 (.80)	16.51 (.65)	19.81 (.78)
100	23 (.90)	19.05 (.75)	22.35 (.88)
132	28 (1.10)	24.13 (.95)	27.43 (1.08)
196	38 (1.50)	34.29 (1.35)	37.59 (1.48)

Note 1: The length dimension is bumper to bumper. The "A" width dimension is inside plastic to inside plastic. The "B" width dimension is lead tip to lead tip.
Note 2: The lead pitch is .635mm (.025″) in all cases.

See Table 2.23 for BQFP component dimensions. See Figure 2.20 for land patterns and Table 2.24 for land pattern dimensions.

The standard QFP in Japan is bumperless. It is controlled by the Electronic Industries Association Japan (EIAJ). They are available in lead pitches of 1mm to 0.5mm (0.0394″ to 0.0197″). Package sizes range from 44 leads to 304 leads.

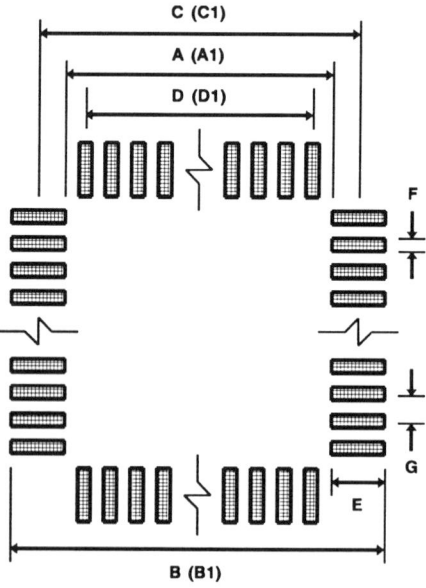

FIGURE 2.20. BQFP (JEDEC) Land Patterns.

Surface Mount Components and Component Packaging   33

TABLE 2.24   BQFP (JEDEC) Land Pattern Dimensions—Metric (Inches)

# Leads	A	B	C	D	E	F
84	17.02 (.670)	21.08 (.830)	19.05 (.750)	12.7 (.500)	2.03 (.080)	.3556 (.014)
100	19.43 (.765)	23.50 (.925)	21.63 (.845)	15.24 (.600)	2.03 (.080)	.3556 (.014)
132	24.64 (.970)	28.70 (1.13)	26.67 (1.05)	20.32 (.800)	2.03 (.080)	.3556 (.014)
196	34.8 (1.37)	38.86 (1.53)	36.83 (1.45)	30.48 (1.20)	2.03 (.080)	.3556 (.014)

Note: The lead pitch (G) is .635mm (0.025") on all BQFPs.

Lead pitch differences are a major concern when designing the PCB. JEDEC uses U.S. units, while EIAJ uses metric units. A JEDEC lead pitch of 0.025" is just that, but the equivalent EIAJ lead pitch is 0.0256" (0.65mm). The same is true for the JEDEC 0.020" lead pitch; the EIAJ equivalent is 0.0197" (0.5mm). A JEDEC component will not fit a land pattern designed for an EIAJ component and vice versa.

See Table 2.25 for QFP component dimensions. See Figure 2.21 for land patterns and Table 2.26 for land pattern dimensions.

Because there are so many leads and because the leads are of the gull-wing style they are susceptible to lead damage. Special care must be taken not to handle the parts in the process. This package style should only be used with placement equipment that has a vision system. The parts should be packaged in matrix trays designed to touch the parts by the bumpers (JEDEC) and not

TABLE 2.25   Square QFP (EIAJ) Component Dimensions—Metric (Inches)

# Leads	A	B	Pitch
80	14.0 (.551)	17.2 (.677)	.65 (.0256)
100	22.0 (.866)	25.19 (.992)	.65 (.0256)
120	27.9 (1.10)	31.24 (1.23)	.8 (.0315)
128	27.9 (1.10)	31.24 (1.23)	.8 (.0315)
144	27.9 (1.10)	31.24 (1.23)	.65 (.0256)
160	27.9 (1.10)	31.24 (1.23)	.5 (.0197)
184	32.0 (1.26)	35.2 (1.386)	.5 (.0197)
208	27.9 (1.10)	31.24 (1.23)	.5 (.0197)

Note: Dimension A is inside plastic width.
Dimension B is lead tip to lead tip width.

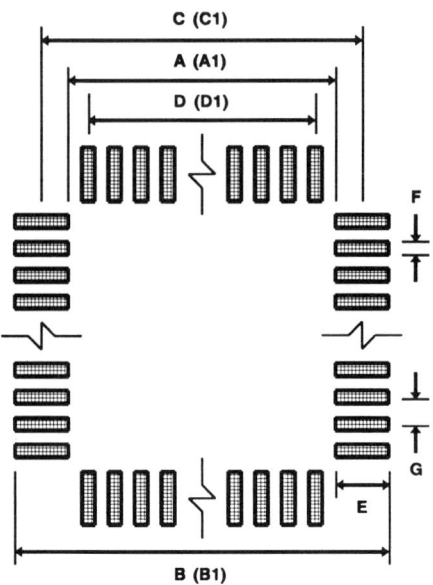

FIGURE 2.21. QFP Land Patterns.

TABLE 2.26 QFP (EIAJ) Land Pattern Dimensions—Metric (Inches)

# Leads	A	B	C	D	E	F	G
80	16.66	20.73	18.69	12.35	2.03	.3556	.65
	(.656)	(.816)	(.736)	(.486)	(.080)	(.014)	(.0256)
100	19.91	23.98	21.94	15.61	2.03	.3556	.65
	(.784)	(.944)	(.864)	(.614)	(.080)	(.014)	(.0256)
120	29.21	33.27	31.24	23.20	2.03	.406	.8
	(1.15)	(1.31)	(1.23)	(.9135)	(.080)	(.016)	(.0315)
128	30.77	34.80	32.77	24.80	2.03	.406	.8
	(1.21)	(1.37)	(1.29)	(.9765)	(.080)	(.016)	(.0315)
144	27.07	31.24	29.21	22.76	2.03	.3556	.65
	(1.07)	(1.23)	(1.15)	(.896)	(.080)	(.014)	(.0256)
160	22.40	25.96	24.13	19.51	1.78	.279	.5
	(.882)	(1.02)	(.95)	(.768)	(.070)	(.011)	(.0197)
184	25.41	28.96	27.18	22.51	1.78	.279	.5
	(1.00)	(1.14)	(1.07)	(.8865)	(.070)	(.011)	(.0197)
208	28.45	31.99	30.226	25.52	1.78	.279	.5
	(1.12)	(1.26)	(1.19)	(1.0)	(.070)	(.011)	(.0197)

TABLE 2.27 TSOP Type I Component Dimensions—Metric (Inches)

# Leads	Length	Width	Height	Pitch
20	16 (.630)	6 (.236)	1.2 (.047)	.5 (.0197)
24	16 (.630)	6 (.236)	1.2 (.047)	.5 (.0197)
28	20 (.787)	8 (.315)	1.2 (.047)	.5 (.0197)
32	20 (.787)	8 (.315)	1.2 (.047)	.5 (.0197)
40	20 (.787)	10 (.393)	1.2 (.047)	.5 (.0197)
48	20 (.787)	12 (.472)	1.2 (.047)	.5 (.0197)

Note: The length dimensions include the leads.

by the leads. QFPs can also be packaged in tape and reel from a few manufacturers.

Thin small outline packages (TSOP) have just recently been introduced into the market. The market needed an integrated circuit package that could be placed as easily as an SOIC, yet had a thinner body package. TSOPs use 20 to 48 gull-wing leads. The pitch is typically 0.5mm (0.0197"). The footprint dimension includes the leads. They are available in two lead configurations: Type I has the leads extending from the narrow ends of the package; Type II has the leads protruding from the wide sides of the package. The standard packaging for TSOPs is matrix trays. See Table 2.27 for Type I and Table 2.28 for Type II component dimensions. See Figure 2.22 for Type I land patterns and Figure 2.23 for Type II land patterns. See Table 2.29 for Type I land pattern dimensions and Table 2.30 for Type II land patterns.

Small outline packages also come in another smaller package known as SSOPs. SSOPs use 8 to 30 gull-wing leads having a pitch of 0.65mm (0.0256").

TABLE 2.28 TSOP Type II Component Dimensions—Metric (Inches)

# Leads	Length	Width (body)	Height (max)	Pitch
20	17.15 (.675)	7.62 (.300)	1.2 (.047)	1.27 (.050)
24	18.41 (.725)	10.16 (.400)	1.2 (.047)	1.27 (.050)
26	17.15 (.675)	7.62 (.300)	1.2 (.047)	1.27 (.050)
28	18.41 (.725)	10.16 (.400)	1.2 (.047)	1.27 (.050)
32	20.96 (.825)	10.16 (.400)	1.2 (.047)	1.27 (.050)
40	18.41 (.725)	10.16 (.400)	1.2 (.047)	.8 (.0315)
44	18.41 (.725)	10.16 (.400)	1.2 (.047)	.8 (.0315)
50	20.96 (.825)	10.16 (.400)	1.2 (.047)	.8 (.0315)

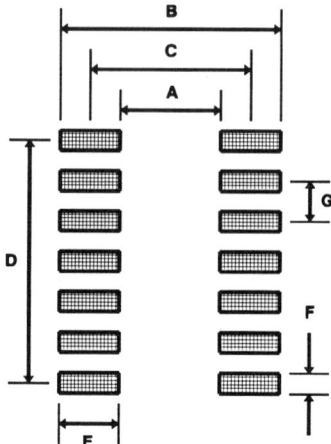

FIGURE 2.22. TSOP Type I Land Patterns.

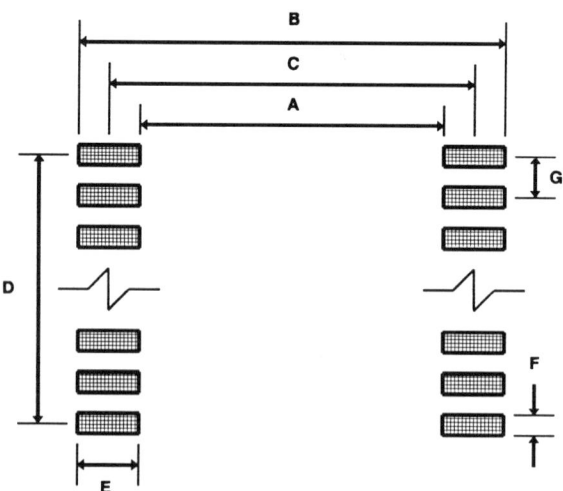

FIGURE 2.23. TSOP Type II Land Patterns..

Surface Mount Components and Component Packaging 37

TABLE 2.29   TSOP Type I Land Pattern Dimensions—Metric (Inches)

# Leads	A	B	C	D	E	F
20	13.69	17.25	15.47	4.50	1.78	.279
	(.539)	(.679)	(.609)	(.177)	(.070)	(.011)
24	13.69	17.25	15.47	5.50	1.78	.279
	(.539)	(.679)	(.609)	(.217)	(.070)	(.011)
28	17.70	21.25	19.48	6.50	1.78	.279
	(.697)	(.837)	(.767)	(.256)	(.070)	(.011)
32	17.70	21.25	19.48	7.50	1.78	.279
	(.697)	(.837)	(.767)	(.295)	(.070)	(.011)
40	17.70	21.25	19.48	9.50	1.78	.279
	(.697)	(.837)	(.767)	(.374)	(.070)	(.011)
48	17.70	21.25	19.48	11.50	1.78	.279
	(.697)	(.837)	(.767)	(.453)	(.070)	(.011)

Note: The lead pitch is .5 (.0197) in all cases.

These are used in credit card memory applications or applications that require very thin profiles.

See Table 2.31 for component dimensions. See Figure 2.24 for land patterns and Table 2.32 for land pattern dimensions.

TapePak is finding its way into the SMT industry in application specific

TABLE 2.30   TSOP Type II Land Pattern Dimensions—Metric (Inches)

# Leads	A	B	C	D	E	F	G
20	6.68	10.74	8.71	15.24	2.0	.66	1.27
	(.263)	(.423)	(.343)	(.600)	(.080)	(.026)	(.050)
24	9.22	13.28	11.25	16.5	2.0	.66	1.27
	(.363)	(.523)	(.443)	(.650)	(.080)	(0.26)	(.050)
26	6.68	10.74	8.71	15.24	2.0	.66	1.27
	(.263)	(.423)	(.343)	(.600)	(.080)	(.026)	(.050)
28	9.22	13.28	11.25	16.5	2.0	.66	1.27
	(.363)	(.523)	(.443)	(.650)	(.080)	(.026)	(.050)
32	9.22	13.28	11.25	19.05	2.0	.66	1.27
	(.363)	(.523)	(.443)	(.750)	(.080)	(.026)	(.050)
40	9.22	13.28	11.25	26.67	.2.0	.66	.8
	(.363)	(.523)	(.443)	(1.05)	(.080)	(.026)	(.0315)
44	9.22	13.28	11.25	26.67	.2.0	.66	.8
	(.363)	(.523)	(.443)	(1.05)	(.080)	(.026)	(.0315)
50	9.22	13.28	11.25	30.48	2.0	.66	.8
	(.363)	(.523)	(.443)	(1.2)	(.080)	(.026)	(.0315)

38   Applied Surface Mount Assembly

TABLE 2.31   SSOP Component Dimensions—Metric (Inches)

# Leads	Body Length	Body Width	Pitch
8	2.99	5.28	.65
	(.118)	(.208)	(.0256)
14	6.19	5.28	.65
	(.244)	(.208)	(.0256)
16	6.19	5.28	.65
	(.244)	(.208)	(.0256)
20	7.18	5.28	.65
	(.283)	(.208)	(.0256)
24	8.18	5.28	.65
	(.322)	(.208)	(.0256)
28	10.2	5.28	.65
	(.400)	(.208)	(.0256)
30	10.2	5.28	.65
	(.400)	(.208)	(.0256)

Note: The overall height for all sizes is 1.84 (.072).

situations. The parts are supplied in a molded carrier ring that keeps the leads flat prior to use. The carrier ring also allows for automated testing. They come in 120 lead to 304 lead sizes. The pitch ranges from 0.63mm to 0.398mm (0.025" to 0.0157"). They are sent to the customer in "coin stack" tubes ready for use on-line. The use of TapePak requires lead forming equipment on-line.

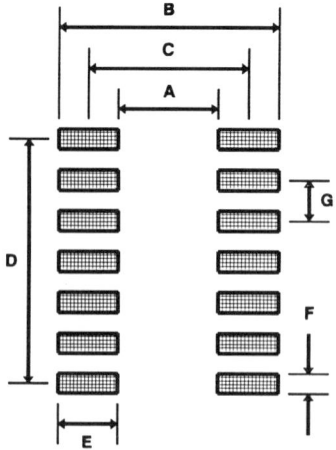

FIGURE 2.24.   SSOP Land Patterns.

TABLE 2.32 SSOP Land Pattern Dimensions—Metric (Inches)

# Leads	A	B	C	D	E	F	G
8	5.33 (.210)	9.4 (.370)	7.37 (.290)	1.95 (.077)	2.03 (.080)	.356 (.014)	.65 (.0256)
14	5.33 (.210)	9.4 (.370)	7.37 (.290)	3.90 (.154)	2.03 (.080)	.356 (.014)	.65 (.0256)
16	5.33 (.210)	9.4 (.370)	7.37 (.290)	4.55 (.179)	2.03 (.080)	.356 (.014)	.65 (.0256)
20	5.33 (.210)	9.4 (.370)	7.37 (.290)	5.85 (.230)	2.03 (.080)	.356 (.014)	.65 (.0256)
24	5.33 (.210)	9.4 (.370)	7.37 (.290)	7.15 (.282)	2.03 (.080)	.356 (.014)	.65 (.0256)
28	5.33 (.210)	9.4 (.370)	7.37 (.290)	8.45 (.333)	2.03 (.080)	.356 (.014)	.65 (.0256)
30	5.33 (.210)	9.4 (.370)	7.37 (.290)	9.10 (.358)	2.03 (.080)	.356 (.014)	.65 (.0256)

This package should be used only in applications that require very high density ICs at a very fine pitch.

2.3 ACTIVE COMPONENTS—CERAMIC

Ceramic chip carriers have been around for more than 15 years, primarily in use by the military. The parts used now in commercial applications in surface mount technology are leadless ceramic chip carriers (LCCC) and leaded ceramic chip carriers (CLCC). The leadless chip carriers, sometimes called LCCs, have solderable castellations for pads. They come in configurations from 16 leads to 44 leads, and in limited use up to 124 leads. They are used in high temperature applications and military requirements. They are shipped in tubes and trays. These packages offer hermeticity and good electrical performance (since all conductor paths are short and essentially the same length). As package sizes grew, problems occurred with cracked solder joints. These packages should not be mounted on epoxy glass substrates because of a mismatch in the coefficient of thermal expansion.

The leaded ceramic chip carrier (CLCC) is a modification of the leadless version. It has filled a much needed vacancy. It is now one of the most dependable packages for higher lead count, reliable circuits. When a higher level of reliability is needed, the CLCC is used instead of the LCCC. Due to the tremendous variation in packages, a land pattern cannot be suggested. Consult the manufacturer for land pattern details.

40   Applied Surface Mount Assembly

FIGURE 2.25.   Surface Mount DIP Switch Example. (Courtesy of American Research Inc.)

2.4 MISCELLANEOUS COMPONENTS

2.4.1 Dual In-line Package (DIP) Switches

Surface mount DIP switches are no more than leaded parts with the leads reformed to accompany the necessities of the surface mount requirements. In certain cases the toggles have been recessed below the top surface of the switch and tape has been placed over the switch to provide a smooth surface for the placement system. They are usually on 2.54mm (0.100″) centers but can be procured on 1.27mm (0.050″) centers. See Figure 2.25 for an example of a surface mount DIP switch.

If the assembly is to be cleaned in solvent make sure the switch is sealed.

See Table 2.33 for component dimensions. See Figure 2.26 for land patterns and Table 2.34 for land pattern dimensions.

2.4.2 Plastic Leaded Chip Carrier (PLCC) Sockets

The original PLCC sockets were leaded. They allowed the use of PLCCs on a leaded PCB. As these sockets improved the problem was reworkability. The

TABLE 2.33   DIP Switch Component Dimensions—Metric (Inches)

# Leads	Length	Width	Pitch
4	5.97 (.235)	7.11 (.280)	2.54 (.100)
8	10.54 (.415)	7.11 (.280)	2.54 (.100)
14	18.67 (.735)	7.11 (.280)	2.54 (.100)
16	21.34 (.840)	7.11 (.280)	2.54 (.100)

Surface Mount Components and Component Packaging 41

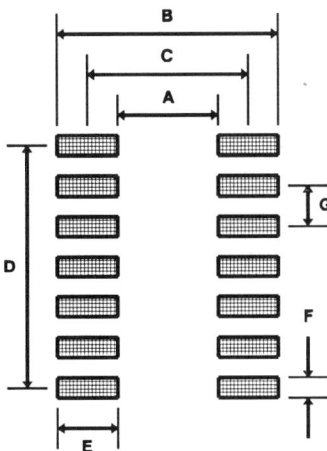

FIGURE 2.26. DIP Switch Land Patterns.

whole socket had to be removed to rework a solder joint. The sockets now available have open areas in the bottom that allow for rework without removal. Also, the footprint of the socket is the same as the PLCC in the socket. This allows the PLCC to be soldered directly to the board in the future, when the mature product does not need the socket. Sockets come in the same size pin counts as PLCCs. See Figure 2.27 for examples of PLCC sockets.

PLCC sockets have the same land pattern layout (Figure 2.19) as the raw components that go in the sockets. The socket can then be deleted from the assembly without redesigning the PCB.

TABLE 2.34 DIP Switch Land Pattern Dimensions—Metric (Inches)

# Leads	A	B	C	D	E	F
4	6.096	11.17	8.636	2.54	.254	.762
	(.240)	(.440)	(.340)	(.100)	(.100)	(.030)
8	6.096	11.17	8.636	7.63	2.54	.762
	(.240)	(.440)	(.340)	(.300)	(.100)	(.030)
14	6.096	11.17	8.636	15.24	2.54	.762
	(.240)	(.440)	(.340)	(.600)	(.100)	(.030)
16	6.096	11.17	8.636	17.78	2.54	.762
	(.240)	(.440)	(.340)	(.700)	(.100)	(.030)

Note: Lead pitch (G) is 2.54 (.100) in all cases.

42   Applied Surface Mount Assembly

FIGURE 2.27.   Surface Mount PLCC Socket Examples. (Courtesy of Amp, Inc.)

2.4.3 Crystals

Crystals are another discrete component that was forced to be packaged in a surface mount package due to customer requirements. The crystal manufacturers merely reformed the leaded parts into surface mount parts. Consult your crystal manufacturer for his package style. There is no established industry standard.

2.5 COMPONENT PACKAGING

All components have to be packaged in some form for shipping and handling. The choices are bulk, tubes, tape and reel (T&R), matrix trays, or custom packaging. For most surface mount components, T&R is becoming the predominant form of packaging. For QFPs the most common packaging form in the United States is matrix trays. The Japanese are starting to package QFPs in T&R. Some unique surface mount components are packaged in tubes because they are hand placed (for example, surface mount D-sub connectors, surface mount switches, and surface mount crystals). Whatever packaging is used it is important that the components are protected during shipping and handling. The packaging also should be compatible with the methods used to process the components during assembly.

2.5.1 Tubes

Tubes are used to supply components, such as SOICs, and PLCCs, in applications where the volumes are too low to justify tape and reel. Rectangular and cylindrical chips and SOTs are not generally packaged in tubes.

Tubes are available in antistatic and conductive materials. There is some concern that the antistatic tubes become ineffective after repeated use.

It is very important that the tube be designed for the component that it is to hold. The wrong tube will not support the component correctly, resulting in damage, especially to the leads.

Attributes
Tubes are a low cost alternative to tape and reel when volumes are low. Suppliers usually charge extra to place components in tape and reel below certain volumes.

Concerns
With the use of tubes comes the potential for putting the tubes in backwards on the placement system (for components that are not symmetrical). Also, for components placed by hand make sure the operator cannot take out more than one part at a time. Do not allow the components to be poured out in a bin and/or on the tabletop.

2.5.2 Tape and Reel

Tape and reel (T&R) consists of three elements: the carrier tape, the cover tape, and the reel. Carrier tapes are made from three materials: paper, plastic, and metal. Paper is used only for resistors. Embossed plastic is used for capacitors, SOTs, and the integrated circuit packages. The use of embossed metal is decreasing in favor of embossed plastic. The carrier tape is made by punching a hole in the tape (paper) or by forming a cavity in the tape (plastic and metal). Sprocket holes are punched on one or both sides. These sprocket holes mate with a drive sprocket on the feeder. The plastic tape is antistatic or conductive. The cover tape, which is also antistatic or conductive, is applied to the top of the carrier tape. Paper tape also requires a cover tape on the bottom.

Carrier tapes are available in seven widths: 8mm, 12mm, 16mm, 24mm, 32mm, 44mm, and 56mm. Tape pitch, which is the distance from the center of one cavity to another, is also important. It varies depending on the tape width. The force required to peel the cover tape from the carrier tape is important. If it is too low the cover tape will fail before use; if it is too high the feeder will not be able to peel the cover tape off to expose the component.

The 8mm and 12mm tapes are available in 7-inch and 13-inch reels, 16mm tapes and larger are only available in 13-inch reels. See Figure 2.28 for an example of tape and reeled components.

Attributes
Tape and reel is the most effective method for introducing components to the placement system. Tape and reel feeders are more reliable than tube feeders. The tape and reel standard, EIA481, is universally accepted.

Concerns
One area of concern is the use of paper reels for the large reels, i.e., 44mm and 56mm. Paper and cardboard are not as rigid as plastic. Another area of concern is the peel strength for the carrier tape. It is still a problem. Some suppliers are not always able to achieve the correct peel strength.

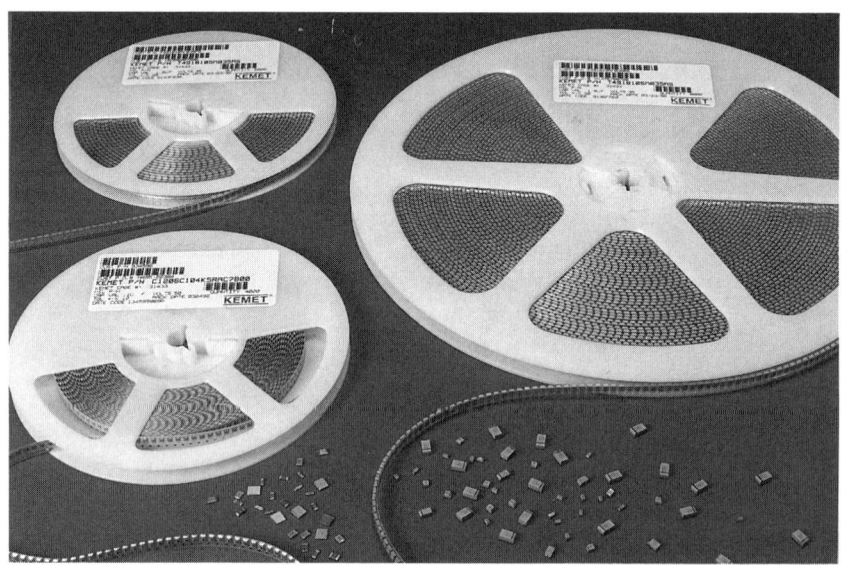

FIGURE 2.28. Surface Mount Tape and Reel Examples. (Courtesy of Kemet Electronics Corp.)

2.5.3 Matrix Trays

Matrix trays were developed out of necessity for the handling of quad flat packs. They hold the components in place without damaging the fragile leads. They are available in two forms of plastic. One can be used to hold the components during the bake cycle (see IPC-SM-786, "Testing and Handling of Surface Mount Plastic Packages Susceptible to Moisture-Induced Cracking"). It is made of high-temperature plastic. The other is a low temperature plastic not suitable for baking. Matrix tray standards can be found in JEDEC Publication 95.

Attributes
In the United States most suppliers package BQFPs in matrix trays. Trays that hold by the body are better than ones that hold by the leads. Bumpered QFPs work great in trays. The trays can be quite expensive, so it is advantageous to set up a program to reuse and/or recycle them. The Japanese tape and reel QFPs now.

Concerns
The major concern over the use of matrix trays is the cost factor. They are more expensive than tape and reel. More United States suppliers will therefore follow the Japanese lead of moving toward tape and reel.

REFERENCES
1. Coombs, Clyde. *Printed Circuit Handbook*. New York: McGraw Hill, 2nd ed, 1979, pp. 2–18, 23–5.
2. EIA-RS-481. "Surface Mount Tape and Reel Packaging." IPC, Lincolnwood, IL.
3. IPC-SM-782. "Surface Mount Land Patterns: Configurations and Design Rules." IPC, Lincolnwood, IL, March 1987.
4. IPC-SM-786. "Testing and Handling of Surface Mount Plastic Packages Susceptible to Moisture Induced Cracking." IPC, Lincolnwood, IL.
5. Prasad, Ray. *Surface Mount Technology*. New York: Van Nostrand Reinhold, 1989.

3
Surface Mount Printed Circuit Boards

GLOSSARY

Array A printed circuit board that has more than one board on the panel. After assembly the individual PCBs are broken out of the panel.
Aspect Ratio The ratio of the circuit board thickness to the smallest hole.
Barrel The cylinder formed by plating through a drilled hole.
Component Hole A hole used for the attachment and electrical connection of component terminations, including pins and wires, to the printed circuit board.
Coplanarity The maximum distance between the lowest and the highest pin when the package rests on a perfectly flat surface. The industry standard is 0.1mm (0.004").
Delamination A separation between any of the layers of a base material or between the laminate and the conductive foil, or both.
Dry-Film Coating material in the form of laminated photosensitive sheets specifically designed for use in the manufacture of PCBs and chemically machined parts. They are resistant to various electroplating and etching processes.
Ductility The extent to which the copper plating can be stretched or elongated before fracture.
Electroless Plating The deposition of conductive material from an autocatalytic reduction of a metal ion on certain catalytic surfaces.
Electroplating The electro-deposition of a metal coating on a conductive object. The object to be plated is placed in an electrolyte and connected to one terminal of a DC voltage source. The metal to be deposited is similarly immersed and connected to the other terminal. Ions of the metal provide transfer to metal as they make up the current flow between the electrodes.

Epoxy Smear Epoxy resin that is deposited onto the surface or edges of the conductive pattern during drilling. Also called **Resin Smear**.

Glass Transition Temperature (Tg) The temperature at which a plastic changes from a rigid, stable state to a soft, unstable state.

Hot Air Leveling (HAL) The technique used to apply solder to the surface of the PCB. Air knives focus the direction of the solder.

Ionic Contamination Test Coupon A small test coupon (section of PCB material) designed to be used for testing for residual salts on the PCB after it has gone through the PCA cleaning process.

Lamninate A product made by bonding together two or more layers of material.

Legend A format of lettering or symbols on the PCB (for example, part number, component locations, and patterns).

Liquid Photoimageable Solder Mask (LPISM) A solder mask that is usually applied using a curtain wave with specialized equipment. It is then cured using photographic imaging equipment. It allows for very thin lines of solder mask in the 0.076mm (0.003″) range. Thus it can be applied between .508mm (0.020″) pitch integrated circuits.

Mask A material applied to enable selective etching, plating, or the application of solder to a PCB.

Microsectioning The preparation of a specimen for the microscopic examination of a specified material, usually by cutting out a cross section, followed by encapsulation, polishing, etching, staining, etc.

Panel The raw laminate material from which a PCB(s) is fabricated.

Plated Through Hole (PTH) A hole that is drilled in the PCB that is copper plated at a subsequent step in the process.

Primary Side The side of the PCB considered to be the top. It is the side that usually has the components that do not go through the wave solder and the side in which through-hole components are inserted.

Screen Printing A process for transferring an image to a surface by forcing suitable media through a stencil screen with a squeegee. Also called **Silk Screening**.

Secondary Side The side of the PCB usually considered to be the bottom. In a typical assembly the rectangular chips are placed on this side.

Solder Bridging The term used to describe the bridge of solder between two or more solder joints.

Solder Mask The protective coating that is applied to one or both sides of the PCB. Several types include: dry film, wet film, LPISM.

Tensile Strength The maximum stress that the metal can withstand under tension.

Tented Vias Plated through-holes that are covered with solder mask.

Test Coupon A portion of a PCB or a panel containing printed circuit coupons, used to determine the acceptability of such boards.

Tooling Holes The general term for holes placed on a printed circuit board, or a panel, for registration and hold-down purposes during the manufacturing process.

Via Hole A plated through-hole used as a through connection, in which there is no intention of inserting a component lead or other reinforcing material.

Wave Soldering The technique used to apply solder to the bottom of the PCB using a wave of molten solder. The PCB moves through the solder on a moving conveyor belt.

3.0 INTRODUCTION

With the advent of surface mount components came the need to decrease the size of the printed circuit board. This required narrow conductors, less than 0.254mm (0.010″), narrow spacing, also less than 0.254mm (0.010″), and smaller plated through-holes, down to 0.508mm (0.020″) in most cases, and below 0.508mm (0.020″) in other very dense applications. Other issues such as solder mask, silk screen, and glass transition temperature (Tg) have also come under pressure to change.

This chapter will review the materials and processes used in the manufacture of surface mount printed circuit boards.

3.1 MATERIAL REQUIREMENTS

The selection of materials for standard multilayer PCBs is important because it affects electrical, thermal, and mechanical quality and reliability. Each material or combination of materials has applications in which they operate very well. They must be evaluated on their own merits for the advantage and/or disadvantage they bring to the table. The PCB design engineer knows that no one single material works for all applications.

The most popular material used for the manufacture of PCBs is still FR-4 (GF). There are numerous other types of materials available such as polyimide (GI), multifunctional epoxies (FR-5), etc. Polyimides have become one of the most significant new materials in the high-end laminate arena. Two properties that are the most critical are coefficient of thermal expansion (CTE) and glass transition temperature (Tg).

The coefficient of thermal expansion is a quantified indicator of how much a particular material expands when it is exposed to heat. At present the average FR-4 PCB has a CTE of 55 (inch/inch/degrees Celcius $\times 10^{-6}$) below Tg. If the resin is not supported it will expand isotropically (in all three dimensions). In FR-4–type PCBs the glass has a lower CTE than the resin and tends to have its expansion restrained in the X and Y directions but not in the Z direction. Below the Tg, the CTE in the Z direction can be as much as ten times compared to the X and Y directions. The obvious worse case result of such expansion is the cracking and/or strain on the plated through-holes (PTH). If up-front reliability testing is not done on each individual PCB design field failures are inevitable. A few of the laminate manufacturers are working on modified epoxy glass laminates to decrease the CTE below 55, to the 45 range.

The glass transition temperature (Tg) is the temperature at which a material changes from a solid into a soft plastic state. At present an average FR-4 laminate has a Tg of about 130°C (266°F). When the glass transition temperature is met several changes take place in many of the properties. These changes can cause many problems in such areas as handling, electrical performance, and processing. In the area of PCB processing this effect can cause smearing while performing the hole drilling operation. Work is under way at many laminate manufacturers to increase the Tg to 160°C (320°F) and higher.

If the PCB design calls for limited Z axis expansion the designer should pick polyimide over FR-4 due to the fact that polyimide has a much higher Tg than FR-4, yet their CTE is similar. Figure 3.1 represents the difference between the Tg of FR-4 and polyimide. Note that the curves are symmetrical, yet they are separated by approximately 100°C (212°F).

The designer has a very complex job of determining just what is needed. The designer must know up front which properties are required based on the application and the processes required to assemble it. Once this is determined the correct substrate material can be selected. Refer to Table 3.1 for property information on various substrate material.

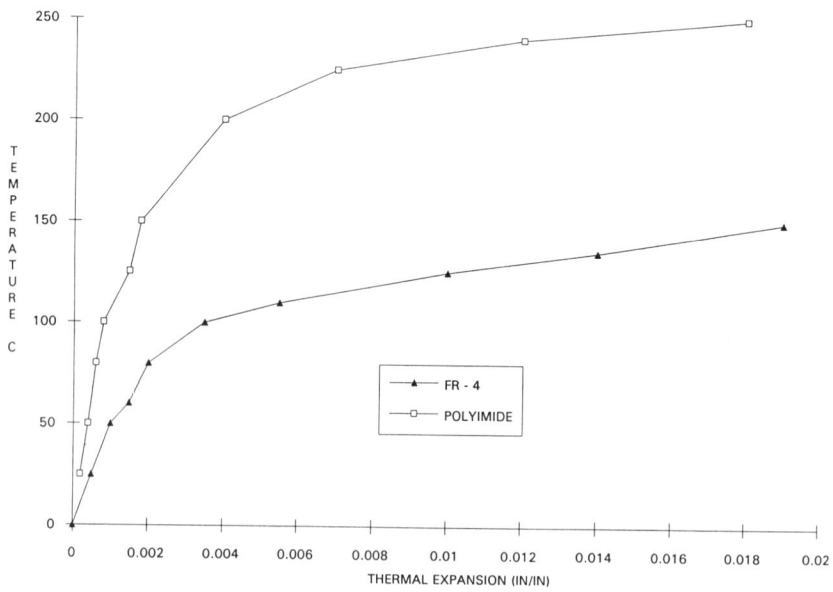

FIGURE 3.1. Comparison of Tg with FR-4 versus Polyimide.

TABLE 3.1 Substrate Material Properties

Type	CTE, PPM/oC −X,Y Axis	CTE, PPM/oC −Z Axis	Thermal Conductivity (Btu/hr/ft/oF)	Tensile Strength Kpsi	Relative Cost
FR-4 Epoxy Glass	12.8–16	189	0.2	40	$2.50 @ 0.060in.
Polyimide Glass	11.7–14.2	60	0.3	50	$6.00 @ 0.060in.
Copper-Clad Invar	5.5–10	16	90–95	60	$6.00 @ 0.025in.
Polyimide Kevlar	3.4–6.7	83	0.15	30	$15.00 @ 0.060in.
Polyimide Quartz	6.0–8.0	6.0–8.0	0.3	—	—

3.2 PHYSICAL REQUIREMENTS

When designing a printed circuit board the design team has to consider the thermal, electrical, and mechanical requirements of the application.

In the area of thermal requirements, PCBs are now being subjected to higher temperatures and longer processing times, both at the laminate manufacturer and at the PCB assembly factory. This is pushing designers to request materials with a higher Tg, i.e., 160°C to 180°C (320°F to 356°F). The higher Tg will allow the PCA process engineer to develop a more robust process. An example would be the liberty to have a larger temperature window in the reflow process. Also, CTE matching is becoming a requirement. A large variation of CTE between mating components can cause cracks and/or separations. New types of reinforcement materials are being evaluated.

Electrical requirements, mostly control of impedance, are becoming very critical in some of the leading edge technology designs. This has pushed laminate suppliers to improve the control of the dielectric constant and tighten the tolerance window.

Mechanical requirements are pushing laminate suppliers as well as the process engineers at the PCB manufacturers. They are obligated to drill smaller holes and etch thinner conductors between narrower spaces. Plated through-holes are getting smaller, <0.33mm (0.013″) while the layer count is increasing, which in some cases requires thicker PCBs >1.58mm (0.062″). See IPC-TR-579, "Round-Robin Evaluation for Small-Diameter Plated Through-Holes." There are studies showing aspect ratios in the 5.5 range passing 400 cycles (per IPC-TR-579) with no failures.

The aspect ratio of the PCB should always be taken into consideration. It is the ratio of the smallest hole to the worst cast board thickness. An aspect ratio above 3.0 should be of concern. On a standard 1.58mm (0.062″) PCB this would mean that there are holes smaller than 0.508mm (0.020″).

3.3 PLATING REQUIREMENTS

In a PCB shop there are many plating processes. The most common being copper, nickel, gold, and tin/lead.

Copper plating is the most common conductive metal used on the surfaces and in the holes of PCBs. For copper plating there are two bath types: electroless and electrolytic. Electroless plating uses no electrical current in the bath, while electrolytic plating uses a charged bath to apply copper to the PCB. The copper plating serves as the conductive path from one point to another. Refer to Table 3.2 for attributes and concerns regarding copper plating.

Nickel plating is typically used on PCBs as an undercoating for gold plating, to form an intermetallic barrier between the copper and gold. A typical thickness for nickel in this application is 100 microinches minimum. Also, there are a few PCB shops that have customers who require plated tin/nickel. This provides a very coplanar surface for applications that need it and also increases serviceability.

Gold plating is used in applications for gold fingers or selective gold. It is deposited using the electrolytic process with one of four plating solutions: acid, alkaline cyanide, alkaline noncyanide, or neutral. Acid gold is the most common. A minimum of 30 microinches is the industry standard thickness for gold fingers.

Tin/lead plating is applied using the folllowing coating method: hot air leveled (HAL), dipped, or electroplated. The tin/lead plating is applied to the conductive areas that will later have components soldered in place. A typical thickness is about 0.025mm (.001″) thick. In the last few years the most common method of applying tin/lead has been HAL. Two types of HAL equipment are available: vertical and horizontal. During the HAL process the PCB goes through a solder bath and then through an area where air is blown across the PCB. The obvious critical variations are the conveyor speed and the fan speed. This process has a potential of a large variation in residual tin/lead thickness. This could be a potential problem (in the PCA process) on

TABLE 3.2   Attributes and Concerns for Copper Plating

Type	Attributes	Concerns
Electroless	Provide conductive path. Compressive residual stress	Low tensile strength Low ductility
Electrolytic	High tensile strength High ductility	Must have conductive path in place

TABLE 3.3 Advantages and Disadvantages of Three Tin/Lead Plating Methods

Type	Attributes	Concerns
Electroplated	Very controlled thickness	Very porous until fused
	Excellent coplanarity	Poor adhesion until fused
Dipped	Low cost equipment	Uncontrolled thickness
	Dense	Non-uniformity
	Good adhesion	
Hot Air Leveled	Dense	Uncontrolled thickness
	Good adhesion	Nonuniformity
	Increased thru-put	Poor coplanarity

lands that a high pin count integrated circuit must be placed. Refer to Table 3.3 for attributes and concerns of tin/lead plating methods.

3.4 SOLDER MASK REQUIREMENTS

The solder masks of today are much different than their predecessors. The original use (in the 1960s) was on the secondary side of the printed wiring board to stop the exorbitant use of solder, which would wick to the traces during wave soldering. As circuitry began to get narrower a second use became apparent. It was used to minimize solder bridging between adjacent traces and/or pads.

In the beginning all solder masks were applied using a printing technique through a tensioned mesh screen. Some of the early solder masks were PC401 and SR1000.

Solder masks are categorized as temporary and permanent. Temporary solder masks are used during the etching processes. The mask is used as an acid resist. It is screened on (wet film) or pressed on (dry film) the copper in the area that is considered to be the circuit. After the acid bath it is washed off leaving the circuitry.

Permanent solder masks are applied to both sides of the completed PCB. These are the solder masks that the PCA factories are familiar with. Permanent solder masks are further broken down into dry film and wet film. Dry film was once thought to be the solution to all solder mask problems, but proved not to be the panacea people thought it would be. For very accurate registration as a solder dam it works well. Dry film is the preferred choice for PCBs that need "tented" vias, which are plated through-holes covered with solder mask. Some of the disadvantages of dry film have always been poor encapsulation of traces, poor lamination over PCBs susceptible to warp and twist, and complex processes. It also causes problems when wave soldering surface mount components and during the solder paste printing process.

Screen printable solder masks are still in wide use in the industry. They probably always will be. But as design specifications require solder masks to provide resolutions of 0.254mm (0.010") or less, screen printable solder masks are reaching their limitations. Cost considerations (yield loss) and mask requirements for fine-pitch surface mount assembly are cultivating the use of liquid photoimageable solder masks (LPISM). LPISMs can be screened on, electrostatically sprayed, or applied through the use of a process called curtain coating. Curtain coating is a process in which the PCB is passed at high speed through a curtain of solder mask. It minimizes waste because the material that does not coat the panels is collected and recirculated. It also allows the control of mask thickness in the range of 0.013mm to 0.038mm (0.0005" to 0.0015"), in addition to complete encapsulation. The biggest disadvantage of this technique is the large up-front capital expenditure.

In printed circuit assembly, solder mask properties that influence cleanability and solderability are thickness, chemical resistance, thermal shock resistance, and resolution. The thickness needs to be restricted to allow a controlled transfer of solder paste volume and prevent shadowing during wave soldering. Chemical resistance of the mask affects which flux is used and how it is removed. Thermal shock resistance is necessary to guarantee the board can handle repeated changes in its environment (reflow and/or wave soldering). Resolution affects solderability basically by the mask's ability to dam between closely spaced traces and/or pads and the ability to cover very narrow traces.

Refer to Table 3.4 for attributes and concerns regarding solder masks.

TABLE 3.4 Attributes and Concerns Regarding Solder Masks

Type	Attributes	Concerns
Liquid Acrylic	Low floor space requirements	Film thickness variations
	Moderate capital equipment costs	No tenting capability
	Low processing time	Moderate adhesion
	Low material cost	Bleeding
	No solvent emission	Poor high-density coverage
Dry Film	Tenting capability	Long process times
	Low set-up time	High material costs
	High-density resolution	Suscpetibility to chipping
	Uniform void free film	Large capital equipment costs
LPISM	High-density resolution	Large capital equipment costs

3.5 SILK SCREEN REQUIREMENTS

Silk screen, also known as the legend, is applied to one side and/or both sides of the printed circuit board. It is applied by flooding the framed silk screen mesh with an epoxy paint. It is then pushed through the mesh onto the PCB in the precise locations. It is important to use a color that contrasts with the solder mask color, which is usually green. Some of the most common colors are white, yellow, and black. See Figure 3.2 for silk screen examples.

Typically, the legend is used as a reference tool to determine the location of a component on the printed circuit board. Two different colors can be used on similar PCBs to make it easier for personnel to distinguish between the two assemblies. Also, the different legend colors can be used to distinguish "family" types. Usually there is *not* a cost increase for different colors, so use it to your advantage. It is important to remember that the application of a legend adds to the cost of the PCB. Determine if the need exists first. Do not always assume that it is needed. Many applications do not need a legend and thus it would be a waste of money.

3.6 PRINTED CIRCUIT BOARD PANELS

Printed circuit board suppliers do not manufacture PCBs, they manufacture panels. A panel is usually made up of several PCBs, but it obviously depends on the size of the end product. Some of the most common panel sizes that are used by PCB suppliers are: 18" × 24", 16" × 21", 18" × 21", 14" × 21". Panels are standardized to stay within the size limitations of the PCB processing equipment. If the PCB is designed to stay within one of these panel sizes less material will be scrapped, resulting in a lower end-item cost. To get the most cost-effective design and the highest quality PCB it is important to know just what the panel limitations of the specific PCB supplier are. Most of the time the PCB is not designed by the PCB manufacturer but by the PCA manufacturer. Consequently you must know your PCB suppliers' process limitations. You then have two choices. You can give the supplier the PCB design and allow them to design it into their optimum panel size or you can do it yourself. The latter is probably not the most ideal situation. Most of the time you will find the PCB supplier would like to handle it. Consult your designated PCB supplier and work out the details.

In the last few years a lot of PCA manufacturers have started processing their assemblies with PCBs in an array form. Using the array form allows the PCB(s) to be processed through the surface mount PCA placement equipment without having the need for a conveyor holding fixture. The array is designed to fit the width of the PCA conveyor system. This requires the PCB

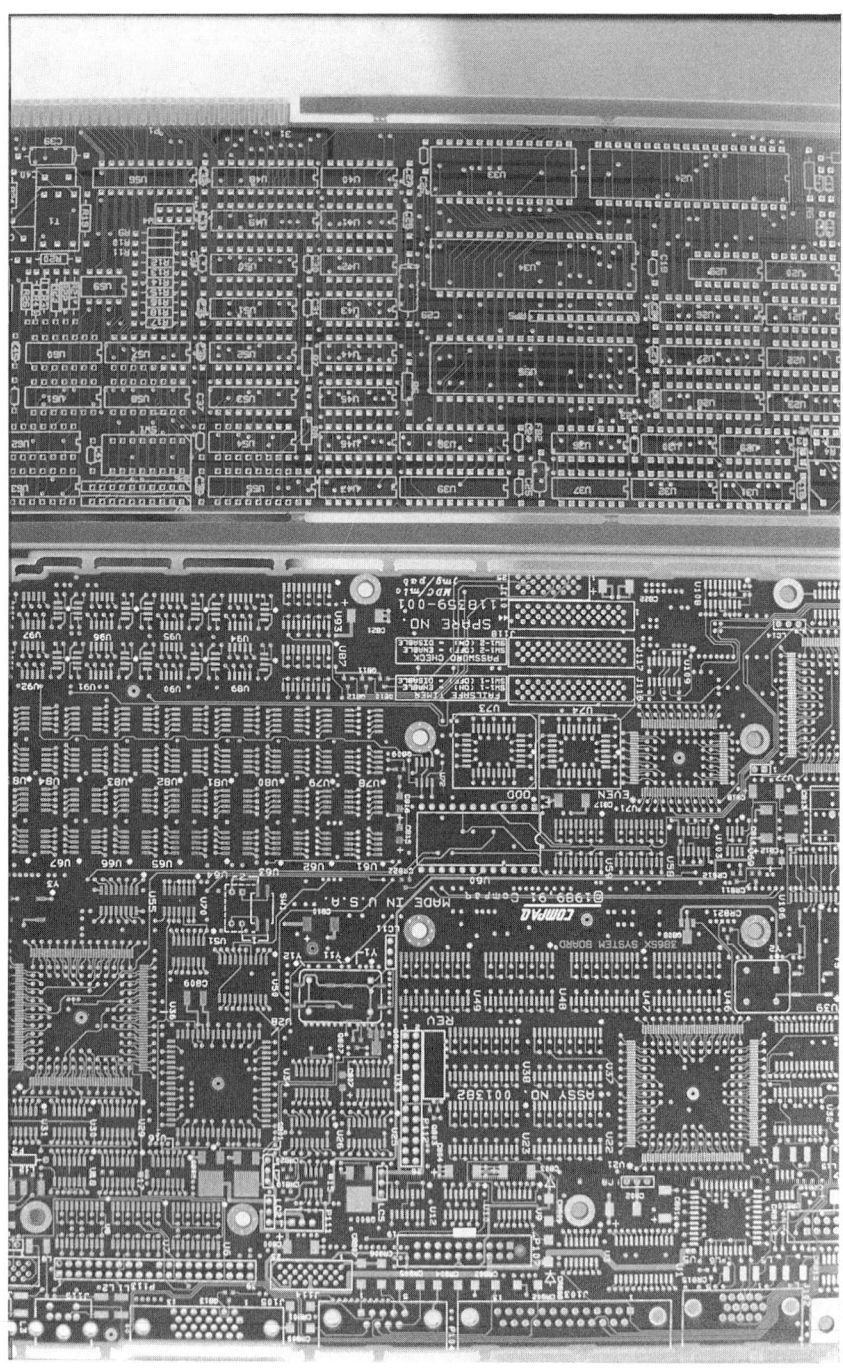

FIGURE 3.2. Examples of Silk Screens. (Courtesy of Praegitzer Circuits.)

to have extra material on the outside edges. Some PCA conveyors require a minimum of 1" on the conveyor sides.

The PCB supplier uses a router to cut an opening between the PCB and the extra material. The width of the cut should be kept as small as possible. A wide cut will make the PCB susceptible to sagging in the PCA process, requiring board stiffeners. Refer to Figure 3.3 for an example of a PCB array.

In the array process the extra material is usually just broken off when the PCA is complete. Also, the extra material protects the edges of the PCB as well as protecting fragile surface mount components that are close to the edge. The extra material can also encompass such things as ionic contamination test coupons, in–out drill monitors, plated holes/pads for cross sections, or any other on-going experiments or tests that might be needed.

FIGURE 3.3. Example of a PCB Array. (Courtesy of Praegitzer Circuits.)

REFERENCES

1. Galvin, Thomas. "Liquid Photoimageable Solder Mask Technologies." *Electronic Packaging and Production*, June 1991, pp. 53–56.
2. IPC-TR-579. "Round-Robin Evaluation for Small-Diameter Plated Through-Holes." IPC, Lincolnwood, IL, September 1988.
3. Jones, Howard. "Basic Overview of PCB Manufacturing." Proceeding of the Nepcon-West Conference, IPC, Lincolnwood, IL, February 1992.
4. Murray, Jerry. "So Many Solder Masks." *Circuits Manufacturing*, February 1990, pp. 56–59.
5. Prasad, Ray. *Surface Mount Technology*. New York: Van Nostrand Reinhold, 1989.

4
Designing for Assembly

GLOSSARY

Breakaway Segments Material connected to a single PCB or array to extend the width and/or length to achieve a standard size.

Breakaway Tabs Tabs of material used to interconnect individual PCBs of an array or breakaway segments. Also, called **Break-off Tabs.**

Fiducial A mark on a PCB used as a target by a vision alignment system.

Fiducial, Global A fiducial target that is used to locate the position of all of the land patterns on a PCB.

Fiducial, Local A fiducial target (or targets) used to locate the position of an individual land pattern on a PCB.

Fiducial Target A PCB artwork feature (or features) that is created in the same process as in the PCB conductive pattern. It provides a common measurable point for component mounting with respect to a land pattern or land patterns.

Finished Hole Size (FHS) The completed hole after drilling and plating. This is the size hole the customer should specify.

Flux A substance used to assist the soldering of metals by removing oxidation and contamination.

Land Pattern The surface area or lands on a PCB on which surface mount components are placed and soldered.

Tooling Hole A nonplated hole used to position the PCB or array on assembly or test fixtures.

4.0 INTRODUCTION

This chapter will focus on the importance of manufacturability. It will review component selection, component packaging, component layout, pad and hole size, conductor size and routing, as well as PCB and array design.

4.1 WHAT IS MANUFACTURABILITY ALL ABOUT?

Design for assembly (DFA) refers to the ability to manufacture printed circuit assemblies (PCA) in the factory with minimal problems and attain relatively high yields. It is directly impacted by the complexity and/or the unique requirements in either product or process. The design engineer and process engineer should minimize the number of possible variations. Assemblies that dramatically deviate from the norm—i.e., combine mixed technologies, use unique parts, require special handling, etc.—cause problems for manufacturing and thus can decrease manufacturability. Designing for assembly is an art. It requires a major effort from the individuals designing the PCA. In today's competitive environment failure to do a good job of DFA can lead to failure of the product due to excessive cost and/or poor quality.

4.2 COMPONENT SELECTION

Component selection plays a major role in how manufacturable the printed circuit assembly is in production. The design engineer/component engineer's first choice should always be to use the components that are presently available in-house (used on previous designs). If the component is going to be new to the production area choose components that use industry standard packaging. The last choice would be to introduce unique components and packaging.

When a component is selected use the following criteria to ensure that a suitable component is used:

1. Packaging—use industry standards (for example, tape and reel (T&R), matrix trays). (See Chapter 2.)
2. Dimensions—the shape must be useable with current placement equipment. (See Chapters 2 and 7.)
3. Terminations—tin/lead alloys are preferred. (See Chapters 2 and 5.)
4. Cleaning—must withstand current and future cleaning applications. (See Chapter 10.)
5. Thermal—must meet the thermal requirements of the process. (See Chapters 8 and 9.)

Surface mount components come in many shapes and sizes. The shape that is chosen is very critical. As stated before it is very advantageous to stay with the industry standards. This makes it easier for the placement equipment to handle and also makes the components more readily available and typically lower cost. Some examples of standards are: resistors and capacitors use

Designing for Assembly   61

TABLE 4.1   Most Common Surface Mount Components

Component Name	Lead Type	Abbreviated Description	Carrier Tape Width
MELF	End terminations	MML41 MML34	8mm or 12mm
Rectangular passive chip	3- or 5-sided terminations	R or C or I	8mm or 12mm
Transistor or diode	Gull-wing	SOT23, SOT89, SOT143	8mm
Cylindrical diode	Plated terminations	SOD80	8mm
Leadless ceramic chip carrier	Castellation	LCCC	16mm, 24mm 32mm, 44mm
Ceramic leaded chip carrier	Gull-wing or J lead	CLCC	16mm, 24mm 32mm, 44mm
Small outline integrated circuit	Gull-wing	SOIC	12mm, 16mm 24mm
Small outline large IC	Gull-wing	SOLIC	16mm, 24mm
Small outline J lead IC	J lead	SOJIC	24mm
Plastic leaded chip carrier	J leads	PLCC	16mm, 24mm 32mm, 44mm 52mm, 68mm
Plastic quad flat pack	Gull-wing	BQFP, QFP	Matrix tray
Ceramic quad flat pack	Gull-wing	CQFP	Matrix tray

1210/1206/0805/0603; transistors use SOT23, SOT89, SOT143, and DPAK; integrated circuits use gull-wing leads and if possible avoid the use of J leads, which are more difficult to inspect and rework. Many times the same components are available in several package styles. The following is a list of package styles that should be avoided: PLCCs with more than 68 pins, MELFs, butt leads, LCCC (leadless ceramic chip carriers), and lead pitches below 0.508mm (0.020″).

4.3 COMPONENT PACKAGING

It is very important to have components that will be used in packaging that is compatible with the placement equipment that will be processing the assembly. See Chapter 2 for details.

4.4 COMPONENT SPACING

Component to component spacing is an important factor for placement, reflow, cleaning, inspection, and rework. Tables 4.2 and 4.3 list basic guide-

TABLE 4.2 Component Type

A	B	C	D	E
0402	1812	SOIC	SOJ	PLCC52
0603	1825	SOLIC	PLCC18	PLCC68
0805	SOT89	SOP	PLCC20	PLCC84
1206	Tantalum	SOM	PLCC28	Connector
1210	Inductor	TSOP	PLCC32	
MLL34		VSOP	PLCC44	
MLL41			QFP	
SOT23			DIP	
SOT143			Axial	
			Radial	

TABLE 4.3 Component Spacing—Land-to-Land, Land-to-Body, or Body-to-Body. Dimensions Are in Inches.

	A	B	C	D	E
A	0.025	0.040	0.050	0.100	0.150
B	0.040	0.040	0.050	0.100	0.150
C	0.050	0.050	0.050	0.100	0.150
D	0.100	0.100	0.100	0.100	0.100
E	0.150	0.150	0.150	0.150	0.150

lines for component spacing. These guidelines may vary depending on the placement and reflow equipment that is used, but they should be viable for most applications. Depending on the component type and orientation, the spacing may be from land to land, land to component body, or component body to component body.

Example: How close can an SOP be to another SOP? Use column "C" in Table 4.2. Then find the dimension that corresponds to "C" in the X and Y positions of Table 4.3. That dimension is 0.050″.

4.5 COMPONENT LAYOUT

Surface mount components can be placed on the printed circuit board on one or both sides. Every effort should be made to design the PCA so that all components are placed on one side of the PCB. A one-sided population provides the best situation for production tooling, testability, and most importantly high yields. Failing this, a two-sided population should adhere to the following requirements:

Primary Side Components
1. Rectangular chips and cylindrical components
   Examples: 0603, 0805, 1206, 1210, 1812, 1825, 2512, 2817, MLL34, MLL41
2. Active discretes
   Examples: SOT23, SOT89, SOT143, DPAK
3. Integrated circuits
   Examples: SOIC, SOLIC, SOMIC, SOJIC, PLCC, BQFP, QFP, CQFP, CLCC

All leaded components should be placed on the primary side. The use of them on the secondary side increases the complexity of the PCA process and typically will decrease yields as well as increase cost.

Secondary Side Components
1. Rectangular chips
   Examples: 0603, 0805, 1206, 1210
2. Active discretes
   Examples: SOT23, SOT89

4.6 PAD AND HOLE SIZE

Plated through-holes (PTHs) are used to interconnect conductors internally and/or externally. They are drilled with various size drills. The most common size holes are in the 0.508mm (0.020″) range, but numerous companies are drilling PTHs below 0.508mm (0.020″) and some below 0.254mm (0.010″). At this point the aspect ratio becomes a big problem. Some companies are experimenting with putting PTHs in lands for PCBs with major space problems. High first-pass yields cannot be achieved due to the PTH pulling the molten solder from the land, resulting in insufficient solder. PTHs can be placed under integrated circuits with a minimum of problems, especially if the PCA is a "no-clean process," otherwise be aware of the potential for flux entrapment. PTHs that are close to the edge of the panel may not fill with solder during the wave soldering process because they interfere with the conveyer.

Table 4.4 outlines the hole requirements for through-hole components. It details the pad stack-ups.

TABLE 4.4 Pad Stack-Ups for Leads Manually Inserted in Plated Through-Holes (Inches)

Lead Size	Drill	Finished Hole Size	Pad	Anti-Pad
0.017–0.021	#57	0.039	0.060	0.085
0.022–0.024	#56	0.043	0.065	0.090
0.025–0.030	#55	0.048	0.070	0.095
0.031–0.033	#54	0.051	0.075	0.095
0.034–0.039	1/16	0.058	0.080	0.105

4.7 CONDUCTOR SIZE AND ROUTING

The printed circuit board industry has gone through a big change in the past few years in the area of conductor and space width. In the mid- to late 1980s 0.010" (10/10) conductors and spaces were considered the limit. Surface mount technology has pushed the industry to the point that 0.008" (8/8) is becoming commonplace, with 0.005" (5/5) just on the horizon as the standard. General recommendations for conductor/space widths are: use 10/10 as often as possible; use 8/8 if conductors must be routed between integrated circuit lands (SOIC, PLCC, etc.); avoid going below 8/8.

4.8 PCB AND ARRAY DESIGN

A list must be developed by manufacturing engineering detailing the "preferred array size" of printed circuit board arrays to be run down each SMT line. It should take into consideration the equipment constraints and the product needs. It would include widths as well as lengths. This information should then be used by the design engineering team to make sure the PCB is useable on the SMT equipment. Tooling hole sizes vary. Two common sizes are: 3.25mm (0.128") and 4mm (0.157"). Two tooling holes are required. They should be on the same axis. Most equipment have the tooling pins located on the front, but some suppliers still locate them on the back of the equipment.

If the PCB manufacturing area uses stiffeners you should leave a minimum area of 12.7mm (0.5"). This area should be clear of all conductors, pads, and components. Some of the PCBs developed today are quite thin and therefore require stiffeners.

4.8.1 Fiducials

Fiducial targets are required on all printed circuit boards, assuming that the surface mount assembly equipment used has optics capability. These targets are used to get very precise image alignment of the component to the PCB during assembly. There are two types of fiducial targets: global (PCB) level and local (component) level. The actual shape and size of the fiducial may vary depending on the equipment supplier. Local fiducials are used with fine-pitch components. With some SMT equipment it is better for the targets to be outside the fine-pitch component, while other equipment uses one target in the center of the land pattern and one target outside the land pattern. The target outside the land pattern may be used for more than one component placement. Follow recommendations provided by the equipment supplier. Placement of global targets is typically on the diagonal at the panel corners at the longest diagonal. This provides the best angular correction. It is not uncommon to place them at each corner. Two local targets placed diagonally to each other is the most common technique. Some manufacturers use three local fiducial targets when it is necessary to provide the most accurate correction for both translational and rotational offsets. For these applications the three local fiducial targets should be in a triangular pattern and located as far apart as possible within the perimeter of the land pattern. On very long arrays (greater than 14″) fiducials may be placed at interim points on the panel as well as the corners. The Surface Mount Equipment Manufacturers Associations' (SMEMA) fiducial mark standard is shown in Figure 4.1. The shape of the fiducial should be a solid filled circle. The minimum diameter of the fiducial target should be 1.02mm (0.040″). The maximum diameter of the fiducial target should be 3mm (0.118″). The fiducial targets on the same PCB should not vary in size by more than 0.025mm (0.001″).

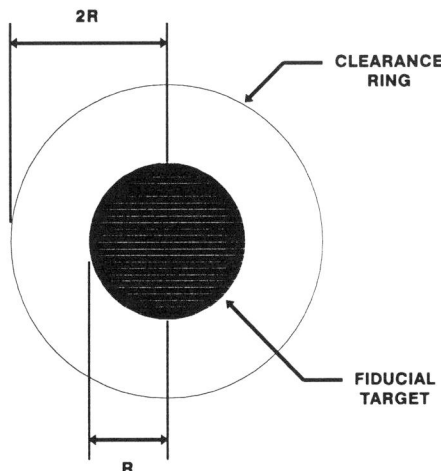

FIGURE 4.1. SMEMA Fiducial Target.

There should always be a clearance area around each fiducial target that is free of any other conductive patterns or markings. The size of the clearance area should have a radius that is at least twice that of the fiducial target and should be concentric with the center of the fiducial target.

The distance from a fiducial target to the edge of the PCB or array should be less than the combination of 4.75mm (0.187″), which is the SMEMA mechanical equipment interface standard, and the fiducial target clearance. The fiducial target shall be bare or covered copper. The covering may be tin, tin/lead plated, or other plating like nickel. For ideal performance the target should have a large contrast in relation to the surrounding PCB base material. The surface of the fiducial target should be flat within 0.0015mm (0.00006″).

4.8.2 Miscellaneous Design Details

To preclude false actuations of conveyor line sensors large cutouts and/or openings should be kept to less than 6.35mm (0.25″) maximum.

On arrays with breakaway segments connecting tabs should be positioned approximately every 25.4mm to 38.1mm (1.0″ to 1.5″) to provide panel support. When trailing and/or leading edge breakaways are required, they should extend the entire width of the panel so that they provide support during the wave and reflow soldering processes. Several techniques for establishing breakaway tabs are used. For flush finished boards, 5.59mm (0.022″) holes are drilled in line with the board edge on 1.27mm (0.050″) centers. This will provide a clean break.

Each of the outside corners of a single PCB or array should be chamfered. The recommended chamfer dimension is a 1.52mm (0.060″) radius or a 45° angle. This will aid the movement of the panel down the conveyor line and through any assembly equipment.

REFERENCES
1. ANSI/IPC-S-815B. "General Requirements for Soldering Electronic Interconnections." IPC, Lincolnwood, IL.
2. Coombs, Clyde. *Printed Circuit Handbook.* New York: McGraw Hill, 2nd ed, 1979.
3. IPC-SM-782. "Surface Mount Land Patterns: Configurations and Design Rules." IPC, Lincolnwood, IL, March 1987.
4. Prasad, Ray. *Surface Mount Technology.* New York: Van Nostrand Reinhold, 1989.
5. SMEMA Standard Fiducial Mark Standard 3.1.
6. SMEMA Mechanical Equipment Interface Standard 1.1.

5
Soldering Materials and Related Issues

GLOSSARY

Activator   A chemical agent added to a flux formulation that promotes the removal of oxidation and contamination.

Alloy   A substance made by melting two or more metals together.

Diffusion   In soldering, the spreading of the molten solder alloy over the base metal surface.

Eutectic Alloy   A unique alloy that has the lowest melting temperature of the two combined metals and a single melting temperature (no plastic range).

Flux   A substance used to assist the soldering of metals by removing oxidation and contamination.

Intermetallic Layer   A compound layer of copper and tin formed at the interface of a copper lead and a tin/lead coating.

Leaching   An event in which one element dissolves into another element above liquidus. Related to SMT it is the dissolution of a precious metal termination coating into the molten tin/lead solder.

Liquidus   The melting temperature of an alloy. If a noneutectic alloy is used it is the highest temperature of the plastic range.

Noneutectic Alloy   An alloy with a melting range. The melting temperature can range from several degrees C to about 30°C.

Poise   A centimeter-gram-second unit of viscosity equal to the viscosity of a fluid that would require a shearing force of one dyne to move a square-centimeter area of either of two parallel layers of fluid one centimeter apart with a velocity of one centimeter per second relative to the other layer with the space between the layers being filled with a fluid.

Solder A metal alloy, with a melting temperature below 450°C (842°F), used for joining two or more base metal surfaces together.

Solder Paste A homogeneous mixture of a powdered metal alloy, a flux system and a viscosity modifier.

Soldering The process of joining two or more similar or certain dissimilar base metal surfaces together using solder.

Solidus The solidification temperature of an alloy. If a noneutectic alloy is used it is the lowest temperature of the plastic range.

Surface Mount Adhesive An acrylic or epoxy adhesive used to attach surface mount components to a PCB.

Vehicle A chemical formulation that acts as a carrier to other materials in a flux or solder paste.

Viscosity The property of a fluid that enables it to develop and maintain an amount of shearing stress dependent upon the velocity of flow and then to offer continued resistance to flow; or the ratio of the tangential frictional force per unit area to the velocity gradient perpendicular to the direction of flow of the fluid.

Thermoplastic A plastic (or adhesive) with molecules that do not cross-link after curing. Subsequent heating will soften the plastic.

Thermoset A plastic (or adhesive) with molecules that cross-link after curing. Subsequent heating will not soften the plastic.

5.0 INTRODUCTION

This chapter will review the various materials used in the SMT assembly process. They include adhesives, fluxes, solders, and solder pastes. The last section of the chapter will review soldering basics. Proper selection, use, and control of these materials are essential to the success of the SMT assembly process.

5.1 SURFACE MOUNT ADHESIVES

Single part adhesives are preferred in this application because they have a much longer working life. Two insulating (nonconductive) adhesive types are commonly used: acrylics, which are thermoplastics, and epoxies, which are thermosets. Both types have a shelf life of approximately two months when stored at room temperature, and up to six months when refrigerated (always follow manufacturers' recommendations for storage). The glass transition temperature (Tg) of the adhesive is an important consideration. The higher the Tg the more difficult rework becomes. The Tg can range from approximately 120°C (248°F) for acrylics to 200°C (392°F) for some epoxies. It is important to select the proper Tg for the application.

The acrylic adhesives can be cured using ultraviolet (UV) light and/or heat. The adhesive must be exposed directly to the UV light to cure; heat should always be applied as well to ensure a complete cure. The epoxies are cured using heat. Curing instructions are supplied by the adhesive manufacturers, and they should be followed closely. For a review of equipment that can be used for curing see section 8.9. When heat curing it is best to use a profile that does not exceed the Tg of the epoxy/glass PCB, which is generally about 125°C (257°F).

The adhesive must have a good wet (uncured) strength to hold components in place prior to curing. Viscosity, which is measured in poise, and is usually noted as centi-poise (CPS), can range from 30,000 CPS for syringe dispensing to 300,000 CPS for stencil printing. When specifying viscosity it is important to note how the measurement was taken, as different equipment will provide different results. Usually noted is the equipment type, spindle type, RPM, reading time, and adhesive temperature (for example, Brookfield HBT Helipath, TF, 2.5 RPM, 2 minutes; 25°C (77°F)). For best results during printing the adhesive temperature should be kept at 25 ($\pm$5)°C and the humidity should be below 65%.

An adhesive color that has a good contrast with the solder mask on the PCB is very useful when inspecting the finished print; red or yellow is common and works quite well.

For additional information on surface mount adhesives, reference IPC-SM-817, "General Requirements for Surface Mount Adhesive."

5.2 SOLDERING FLUX

Flux, which is a chemical mixture, provides two services: it prepares metal surfaces for soldering by removing oxides and other contamination, and it protects the metal surfaces from further oxidation until the soldering process is complete. The selection and use of a flux depends on the end-user application (consumer or high reliability products), the surfaces to be soldered (tin, tin/lead, copper, etc.), and the process (reflow or wave soldering). There are two contradictory functions for a flux: It should be inactive (noncorrosive) at room temperature, but active enough at soldering temperatures to properly prepare the surfaces to be soldered.

5.2.1 Flux Classification

Four flux groups are in use today: rosin (R, RMA, RA, RSA), water soluble (WS), synthetic activated (SA), and low solids (LS), also referred to as no-clean (see section 5.2.2 for details). Table 5.1 shows the solderability process window of the four flux groups. In general, this table indicates that fluxes with low levels of activity have less process capability. Simply stated, when type R, RMA, and LS fluxes are used the base metal surfaces must be

TABLE 5.1   Flux Process Window

Flux Type	Process Window
R	Very Narrow
RMA	Narrow
RA	Wide
RSA	Wide
WS	Wide
SA	Wide
LS	Very Narrow

very solderable. The more active flux types—RA, RSA, WS, and SA—provide an improved process capability because they can remove oxidation and contamination better. The down side to a more active flux is concern about reliability due to corrosion, so cleaning becomes much more important. Flux formulations and activity levels vary, so it is very important to understand what type of flux is being used. Activators in a flux are usually halides (chlorides and bromides) or organic acids.

Several industry and government groups have developed specifications to help classify flux chemistries. The most important of these specifications, outside of military applications, is IPC-SF-818, "General Requirements for Electronic Soldering Fluxes (ANSI/J-STD-004 will supersede IPC-SF-818)." This specification classifies flux according to specific criteria based on flux activity testing and surface insulation resistance testing.

Flux activity is classified as low activity (Type L), moderate activity (Type M), and high activity (Type H), based upon the results of three tests: copper mirror, silver chromate, and corrosion. The copper mirror test evaluates the ability of the flux to remove copper from a slide covered with a very thin layer of vacuum-deposited copper. The silver chromate paper test is performed to ascertain if chlorides or bromides are present in the flux. The copper coupon corrosion test examines the corrosive capability of a flux when it is subjected to severe environmental conditions. Examples of Type L fluxes are all R and LS fluxes, most RMA fluxes, and some RA and WS fluxes. Examples of Type M fluxes are some RMA, WS, and SA fluxes and most RA fluxes. Examples of type H fluxes are some WS and SA fluxes and all RSA fluxes.

Surface insulation resistance (SIR) testing, which measures the impact of flux residue on the insulation resistance of the PCB, has three classifications: 1, 2, and 3. They are based on test storage conditions of 7 days at 50°C (122°F) and 90% relative humidity (Class 1 and 2) or 7 days at 85°C (185°F) and 85% relative humidity (Class 3). The flux residue must measure at least 100 megohms in a cleaned or uncleaned condition, depending on the classification.

A typical RMA flux could have a classification of L3CN, which means it is a low (L) activity flux that passes the cleaned (C) and not cleaned (N) requirements for Class 3 SIR. An important item to consider about flux is that the activity level can vary greatly from one formulation to another, even though two fluxes have the same industry standard nomenclature, such as RMA.

For more information refer to IPC-SF-818, "General Requirements for Electronic Soldering Fluxes. (ANSI/J-STD-004 will supersede IPC-SF-818)." Another flux specification, Bellcore TA-NWT-000078 "General Physical Design Requirements for Telecommunication Products and Equipment," is used by the telecommunication industry.

5.2.2 Flux Groups

Rosin flux is currently divided into four groups: rosin (R), rosin mildly activated (RMA), rosin activated (RA), and rosin super activated (RSA). The R flux is generally considered too inactive for electronics soldering. All other rosin fluxes have added activators. The RMA flux is by far the most common type in use today. An RMA flux has a low to moderate activity level, and is suitable for most soldering applications. Most RMA fluxes are not corrosive or conductive after the soldering process. Difficult to solder materials may require the use of the more active RA and RSA fluxes. Rosin has several noteworthy attributes. The primary function of rosin is as a vehicle, but it is also a good insulator and will serve as an encapsulant at room temperature, preventing the movement of flux activators. Depending on the application, R and RMA flux residue can be left on the PCA or removed. All RA and RSA flux residue must be removed from the PCA after the soldering process.

With the move away from solvent cleaning, water soluble (WS) flux is becoming more common and popular. Most WS fluxes are more active than an RMA flux. Because it is more active, a WS flux should promote better solderability on difficult to solder surfaces. A WS flux is often referred to as an organic acid (OA) flux, which is not entirely correct. Some WS fluxes do not use an organic acid. The residue from a WS flux is corrosive and must be removed from the PCA after the soldering process.

A synthetic activated flux is designed to be soluble in chlorofluorocarbon (CFC) solvents. These fluxes are usually quite corrosive, and they should be removed from the PCA after the soldering process. Since the use of CFC-type solvents is being phased out, the use of SA fluxes will also be phased out.

Because of concerns about cleaning, the industry has started to take a serious look at LS fluxes. An LS flux is designed to leave a noncorrosive residue on the PCA. However, the amount of activator in an LS flux is minimal and solderability may be a problem. The solderability of the PCB and components is very critical when an LS flux is used. A controlled atmosphere (inert) may also be required.

5.2.3 Flux Activation

Flux has two attributes that affect the soldering process. To properly remove oxides it is important to understand the activation temperature and activation time of the flux. A common mistake is to use a time/temperature profile that consumes the activator too soon. This is especially true in reflow soldering. On the other hand, it is important to have the flux active long enough to remove the oxides from the PCB, component leads, and in the case of reflow soldering, the solder paste powder. Ideally for reflow soldering, the last of the activator would be consumed just as the solder begins to melt. An acceptable activation time for most flux material is a minimum of 30 seconds and a maximum of 120 seconds. For standard tin/lead soldering the flux usually becomes active at 110°C to 120°C (230°F to 248°F). Consult with the supplier for recommendations concerning their material.

5.2.4 Solder Paste Flux

Flux in solder paste is part of a chemical system. This system can be broken down into four main elements: vehicle (rosins/resins), activator, solvent, and rheological control agents. Rosins/resins, either natural or synthetic, are used to control viscosity, slump, and tackiness. They can also add to the activity of the flux. The solvent is used to dissolve the rosins/resins and activators. Rheological control agents are added primarily to keep the solder powder from separating from the paste mixture.

5.3 SOLDER

Solder provides the mechanical and electrical connection between the component and the PCB. The surface mount solder joint is called upon to provide all of the mechanical strength in terms of attaching the component to the PCB, as opposed to insertion-technology where much of the mechanical strength comes from the lead being inserted through the PCB. Surface mount solder joint strength and reliability is directly related to land pattern design.

Solder alloys used for electronic assembly are referred to as soft alloys or soft solders. These alloys have a melting temperature below 450°C (842°F). The active metal in the majority of solder alloys is tin. Tin is valuable because it enhances the wetting action. During soldering the tin has a reaction with most base metal surfaces, which results in the formation of intermetallic compounds. These intermetallic compound layers are strong but brittle. Solder alloys and base metal surfaces are made up of unary, binary, and ternary systems. A unary system is comprised of a single metal, such as tin. A binary system is comprised of an alloy of two metals, such as tin/lead. A ternary system is comprised of an alloy of three metals, such as tin/lead/silver.

Soft solders have low mechanical strength when compared to metals with a higher melting temperature. When a soft solder is under stress it will move, which is referred to as creep. Unfortunately, creep occurs at a moderately low temperature in soft solders. The soldering process occurs when a solder alloy joins two similar or certain dissimilar base metal surfaces together by diffusion (wetting). Soldering does not require the melting of the base metals, only the joining metal must be melted. Section 5.5 reviews the basics of the soldering process. Phase diagrams, which have been developed for most alloys, including all of the common soldering alloys, are used to explain various metal or alloy characteristics. A good review of tin/lead phase diagrams can be found in the book *Solders and Soldering* by Howard Manko.

The solder alloy can be eutectic or noneutectic in nature. A eutectic solder is unique because it has a single melting point, and the melting temperature is the lowest of the combined alloys. This is an advantage over alloys that are noneutectic, primarily because the noneutectic alloys have a plastic temperature range. Disturbed solder joints can result from movement or vibration of the alloy while it is in the plastic range. It also allows the lowest soldering temperature to be used. For example, tin/lead alloys dominate electronic soldering. Tin melts at 231°C (448°F), while lead melts at 327°C (621°F). The eutectic alloy for tin/lead is 63Sn/37Pb (63% tin and 37% lead). The melting temperature (liquidus) of 63Sn/37Pb is 183°C (361°F), which is much lower than the individual melting temperatures of either tin or lead. Another common tin/lead alloy is 60Sn/40Pb, which is a noneutectic composition. Its melting temperature (liquidus) is 189°C (372°F) and its solidification temperature (solidus) is 183°C (361°F), so it has a plastic or pasty range of 6°C (11°F).

5.3.1 Wave Soldering Alloys

As noted earlier, solder alloys that are eutectic or near eutectic are preferred, which is why the 63Sn/37Pb and 60Sn/40Pb alloys are used almost exclusively for wave soldering. General information about these solder alloys is provided below.

63Sn/37Pb (tin/lead) This is the most common solder alloy. It has good mechanical, electrical, and thermal properties and works well when soldering to copper, nickel, tin, and tin/lead surfaces. As noted before, this eutectic alloy has a liquidus temperature of 183°C (361°F).

60Sn/40Pb (tin/lead) This alloy can be used in the same applications as the 63Sn/37Pb alloy, but because it is a near-eutectic alloy it is not as popular as the 63Sn/37Pb alloy. The liquidus temperature is 189°C (372°F) and the solidus temperature is 183°C (361°F).

The solder pot temperature range is from 240°C to 260°C (464°F to 500°F), but the lower temperature is preferred to limit the amount of preheat required (see section 9.4). Below 240°C (464°F) the flow characteristics of the molten solder become unsuitable for circulation and pumping in the solder pot. The lower solder temperature also helps decrease dross (oxidized solder) production.

Solder contamination is a major concern. Specification ANSI/J-STD-006, "General Requirements and Test Methods for Soft Solder Alloys" (formerly QQ-S-571), outlines specific allowable levels of contamination in solder alloys. Contamination levels of aluminum and copper are of special concern because above a certain level (0.004% for aluminum and 0.2% for copper) their presence can diminish solder joint quality. Solder pot contamination occurs because of the repeated use of the same solder. To prevent problems it is a good idea to have the solder evaluated, per ANSI/J-STD-006 or QQ-S-571, at least once each month.

5.3.2 Reflow Soldering Alloys

As with wave soldering, solder alloys that are eutectic or near eutectic are preferred for reflow soldering. Two alloys, 63Sn/37Pb (tin/lead) and 62Sn/36Pb/2Ag (tin/lead/silver), are the most common solder alloys available for reflow soldering. Information about these and other solder alloys is given below.

63Sn/37Pb (tin/lead) This is the most common solder paste alloy. It has good mechanical, electrical, and thermal properties and works well when soldering to copper, nickel, tin, and tin/lead surfaces. This eutectic alloy has a liquidus temperature of 183°C. The reflow temperature range is 208°C to 223°C (406°F to 433°F).

62Sn/36Pb/2Ag (tin/lead/silver) This alloy is used when silver is present in one or both of the surfaces to be soldered. The addition of 2% silver helps limit silver migration (leaching) during reflow and slightly improves shear strength. Soldering applications include copper, nickel, tin, and tin/lead. This near-eutectic alloy has a liquidus temperature range of 179°C to 189°C (354°F to 372°F). (The eutectic alloy is actually 62.5Sn/36.1Pb/1.4Ag.) The reflow temperature is 204°C to 229°C (399°F to 444°F).

60Sn/40Pb (tin/lead) This alloy can be used in the same applications as

63Sn/37Pb, but because it is a near-eutectic alloy it is not as popular as 63Sn/37Pb and its use and availability is limited. The liquidus temperature range is 183°C to 189°C (361°F to 372°F), and the reflow temperature is 208°C to 223°C (406°F to 433°F).

Other solder alloys would be considered if a step soldering process were to be used. Low-temperature alloys usually contain bismuth (Bi) or indium (In), such as 43Sn/43Pb/14Bi or 50Sn/50In. High-temperature alloys usually contain higher amounts of lead, such as 10Sn/90Pb.

There are four characteristics that must be clearly understood about the solder alloy being used: melting temperature, reflow temperature, time above melting temperature, and cooling rate.

Reflow temperature is generally considered to be 25°C to 40°C (77°F to 104°F) above the lowest melting temperature of the alloy. It is important to achieve this temperature, which allows the solder to wet the base metal surfaces properly. The time above the melting temperature, typically 20 to 60 seconds, is also significant because it allows the solder enough time to properly wet the base metal surfaces.

Cooling also affects the final strength and integrity of the solder joint. In general, solder joints that are cooled at a reasonable rate achieve a small, fine-grain structure. This grain structure provides a stronger, more reliable solder joint. Cool-down rates of 1 to 2°C/second (1.8 to 3.6°F/second) are preferred; however, cooling rates up to 5°C/second (9°F/second) have been used.

For additional information on solder paste alloys reference IPC-SF-819, "General Requirements and Test Methods for Electronic Grade Solder Paste" (superseded by ANSI/J-STD-005).

5.4 SOLDER PASTE

Although solder paste has been used for almost 30 years for electronic assembly, it is a relatively new material to most manufacturing people. Solder paste is a complex mixture of five elements: metal powder, vehicle (rosins/resins), activator, solvent, and rheological control agents. IPC-SP-819, "General Requirements and Test Methods for Electronic Grade Solder Pastes" (superseded by ANSI/J-STD-005), should be used as a reference for additional information.

5.4.1 Solder Paste Characteristics

Solder paste has many variables that can affect the printing process. They include viscosity, slump, tack time, exposure time, percent metal, powder size, and powder shape. In the past it was common to have the user specify certain aspects of the solder paste such as viscosity, percent metal, and powder shape. Today, however, the suppliers have fine-tuned their formulations to the extent that it is better to tell the supplier about the printing requirements and let them recommend the correct formulations.

Viscosity is very important and also very misunderstood, at least when it comes to measuring it. Viscosity is measured in poise, and is usually noted as centi-poise (CPS) or kilo-centi-poise (KCPS). There are two types of viscometers used for viscosity measurement: Brookfield and Malcom. The Brookfield viscometer is considered the industry standard. The Malcom viscometer is interesting because it measures dynamic viscosity as opposed to static viscosity, which the Brookfield measures. This means that the Malcom viscometer measures the viscosity under shear, which can be an advantage because printing is achieved by shearing the material. Although direct correlation is difficult, Brookfield viscosity will be approximately 3 times more than Malcom viscosity, (in other words, a viscosity of 900,000 CPS Brookfield would be approximately 300,000 CPS Malcom). It is very important to stabilize the solder paste for at least two hours at 25°C (77°F) prior to measurement. The measurement should be within ±10% of the viscosity measured by the supplier. When specifying viscosity it must be noted how the measurement was taken because different equipment will yield different results. Usually noted is the viscosity, the equipment type, spindle type, RPM, reading time, and paste temperature—for example, 950,000 CPS, Brookfield RVTD Helipath, TF, 5 RPM, 2 minutes, 25°C (77°F). It is a very good idea to consult with the solder paste supplier and follow their recommendations for viscosity measurement. Viscosity for screen printing ranges from 500,000 CPS to 850,000 CPS, and for stencil printing the range is 850,000 CPS to 1,300,000 CPS. The temperature of the solder paste should be kept at 25 (±5)°C (77 (±9)°F) during printing, for each 1°C (1.8°F) above this temperature the viscosity will decrease by 10,000 to 20,000 CPS.

Slump is the result of a solder paste that cannot hold its printed shape, and is identified by a decrease in height and an increase in length and width of the solder paste deposit. Slump can be further characterized as cold slump and hot slump. Cold slump occurs at room temperature, while hot slump occurs during reflow. Hot slump has the potential to be the most damaging of the two, as excessive slump results in solder bridging.

Tack time is usually associated with the placement process, but it is also important to the printing process. If the solder paste is not tacky it will not adhere to the lands during printing. Several methods have been developed for measuring tackiness. They range from simply placing a component in solder paste and turning the sample upside down, and timing how long it takes for the component to fall off, to more complex testers developed specifically for this purpose.

The percentage of metal, by weight, has a direct effect on viscosity. The percentage of metal in a printable solder paste varies from about 85% for the low point to about 92% for the high point. The most common percent range is 88% to 90% for solder pastes used in printing.

Powder shape affects printing and oxidation control. A sphere is the best powder shape because it has the least surface area. Spherical powder will pass through a screen or stencil opening better than any other shape. A sphere has the least surface area of any shape, which limits the amount of oxidation that can occur. According to IPC-SP-819, 100% of the powder should have a length to width aspect ratio of 1.5 to 1. However, tests have indicated that this may not be possible, at least not at an acceptable cost. It appears that 97% to 98% of the powder can meet this requirement, while the remaining 2% to 3% has a 2 to 1 aspect ratio. This small percentage of irregular shapes does not appear to have a detrimental effect on the printing or reflowing process. Figure 5.1 is an example of good, spherical powder.

Powder size is divided into four groups according to IPC-SP-819. This grouping is based on Tyler (or ASTM) standard sieves. The four types are listed in Table 5.2.

FIGURE 5.1.   Good Solder Paste Powder.

TABLE 5.2 IPC-SP-819 Powder Size Classification

Description	Percent	Powder Size
Type 1	1% or less	150 microns
	80% minimum	150–75 microns
	10% maximum	20 microns
Type 2	1% or less	75 microns
	80% minimum	75–45 microns
	10% maximum	20 microns
Type 3	1% or less	45 microns
	80% minimum	45–20 microns
	10% maximum	20 microns
Type 4	1% or less	38 microns
	80% minimum	38–20 microns
	10% maximum	20 microns

Type 1 powder is normally not used because the powder is too large. Type 2, which is the most common powder, is used for all standard, non–fine-pitch applications. Type 3 and 4 are used for fine-pitch applications. A common method used to indicate powder size, other than using the IPC-type classification, is to list the sieve sizes. A Type 2 powder would be noted as follows: −200 (75 microns) +325 (45 microns). The minus sign means this powder size will pass through the sieve, while the plus sign means this powder size will not pass through the sieve. In general, this indicates that a minimum of 80% of the powder will be between 75 and 45 microns (Type 2 powder).

Good printability will result if the correct solder paste for the application is selected and the variables of viscosity, slump, tack time, exposure time, percent metal, powder size, and powder shape are properly controlled.

5.5 SOLDERING BASICS

The process of soldering has several fundamental components that should be clearly understood. This section is not intended to provide a detailed explanation of soldering. It is an attempt to introduce the user to soldering fundamentals. Electronic soldering is the process of bonding two similar or certain dissimilar base metals (the component lead and the PCB land) together, using a soft alloy, to form an electrical and mechanical interface.

An important aspect of soldering is wetting, which is the spreading of the molten solder over the base metal surfaces. Also, by definition, the wetting process is accomplished with one solid phase (the base metal surfaces) and one liquid phase (the molten solder). It is critical to note that just because solder has covered the base metal surfaces does not mean that the solder has

Soldering Materials and Related Issues   79

properly wetted these surfaces. Oxidation and contamination on the base metal surface will inhibit proper wetting and prevent a good metallic bond from forming. An intermetallic compound layer, which can form between the base metal alloy and the base metal coating or the base metal coating and the solder, can have a positive impact on reliability in some cases, and in other cases it can have a negative impact on solderability and reliability.

An important attribute of good wetting is the contact angle, which is the angle the molten solder makes with the base metal surface. The smaller the angle, the better the wetting. Contact angles of 90° and less are acceptable, contact angles greater than 90° are unacceptable. To illustrate this point refer to Figure 5.2, which shows three contact angle conditions: less than 90°, equal to 90°, and greater than 90°. Figure 5.3 illustrates these three conditions on a rectangular (resistor or capacitor) component.

Problems with the solder or base metal surface will result in a nonwetting or dewetting condition. Nonwetting occurs when the solder will not spread

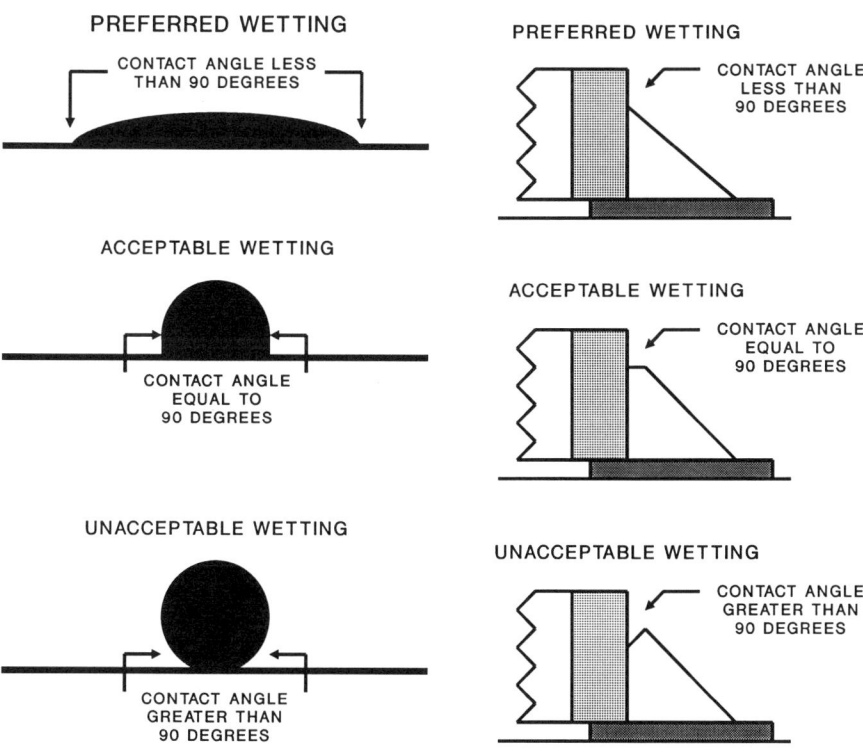

FIGURE 5.2. Solder Contact Angle.

FIGURE 5.3. Solder Contact Angle with Component.

over and bond to the base metal surface. Dewetting occurs when the solder sporadically spreads over and bonds to the base metal surface. Nonwetting is usually an easy condition to detect; dewetting is a more difficult condition to observe.

Nonwetting and dewetting are caused by surface oxidation, surface contamination, or intermetallic compounds. If the base metal surface has too much oxidation or contamination, the flux may not be able to remove them. This will not allow the solder to wet the base metal surface properly. If the coating on the base metal is too thin the intermetallic compound layer will be exposed during the soldering process. Intermetallic compounds have very poor solderability and will not wet with solder. As a general rule, a minimum coating thickness of 300 microinches is required to prevent soldering problems caused by intermetallic compounds.

5.6 COMPONENT SURFACE FINISH

Component leads are typically made from copper and copper alloys. Other alloys, such as Kovar (53% iron, 29% nickel, 17% cobalt) and Alloy 42 (42% nickel, 58% iron), are also used. Copper is used because of its better thermal conductivity. Also, Kovar and Alloy 42 are not as ductile as copper, and therefore are not as well suited to lead forming, especially for J leads.

Component lead finish is an important factor that will affect solderability and reliability. Three methods are used to coat the base metal surfaces: plating, plating and fusing, and hot coating. There is debate within the industry as to which process provides the best solderability, especially long term. Plating provides a more uniform, but porous surface. The porous surface may promote oxidation. If the plated surface is fused by the supplier solderability may be affected by the formation of copper-tin intermetallic compounds. Hot coating provides a thick but uneven surface finish. During application the hot coating tends to pull away from the lead edge, producing an insufficient surface coating condition at the edge. Regardless of the application method, a final surface finish thickness of at least 300 microinches is recommended.

Nonmolded rectangular and cylindrical components may suffer from leaching of the termination, which occurs during the soldering process. Component terminations are constructed by first applying a precious metal adhesion layer, usually silver, to the ceramic. The final coating layer is then applied to the adhesion layer. During the reflow or wave soldering process the adhesion layer is vulnerable to dissolution in the tin. This problem can be prevented by the addition of a barrier layer between the adhesion layer and the outer coating. Nickel, which has one of the lowest dissolution rates in tin/lead solder, is used for the barrier layer. The nickel barrier, which is about

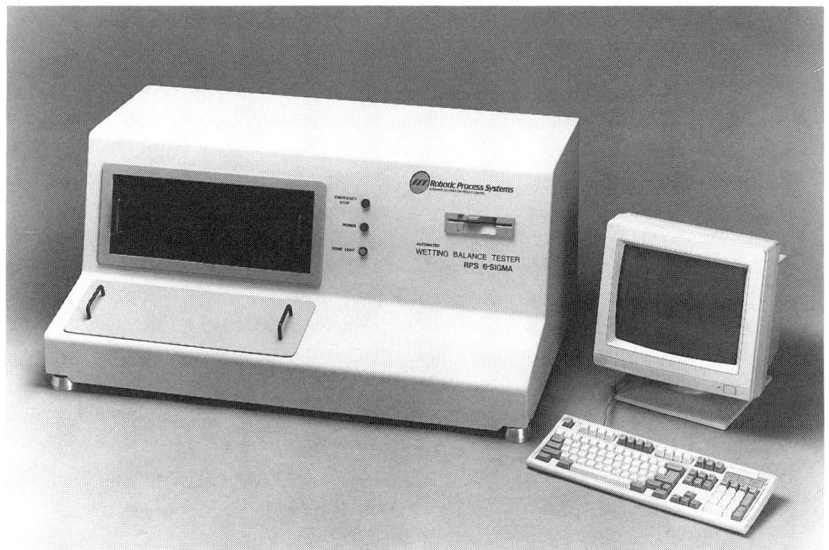

FIGURE 5.4. Wetting Balance Test System.

50 microinches thick, prevents the silver adhesion layer from dissolving into the tin. Another, but less effective, method can also be used to prevent leaching, but only in the reflow soldering process. A solder paste alloy containing 2% silver (62Sn/36Pb/2Ag) is used instead of the standard 63Sn/37Pb alloy. The tin's affinity for silver is satisfied by the silver in the solder paste, and not by the silver in the termination.

Component solderability testing can be helpful in preventing solderability problems during production. ANSI/J-STD-0003, "Component Solderability Test Methods," should be used as a reference. An example of a component solderability tester is shown in Figure 5.4.

5.7 SUBSTRATE SURFACE FINISH

The final PCB surface coating is also applied by plating and fusing or hot coating. A plated surface finish is created by plating tin/lead onto the copper lands and then fusing them using an infrared oven or hot oil bath. A hot coating is applied using a horizontal or vertical hot air leveling (HAL) system. The HAL process tends to produce a very uneven surface finish. In addition to solderability problems, this may also cause coplanarity problems. In either case, a minimum coating thickness of 300 microinches is recommended to ensure good long-term solderability.

REFERENCES

1. ANSI/J-STD-006. "General Requirements and Test Methods for Soft Solder Alloys," IPC, Lincolnwood, IL, October 1992.
2. Hwang, Jennie S. *Solder Paste in Electronics Packaging.* New York: Van Nostrand Reinhold, 1989.
3. IPC-SF-818. "General Requirements for Electronic Soldering Fluxes." IPC, Lincolnwood, IL, March 1988.
4. IPC-SP-819. "General Requirements and Test Methods for Electronic Grade Solder Paste." IPC, Lincolnwood, IL, October 1988.
5. IPC-SM-817. "General Requirements for Surface Mount Adhesive," IPC, Lincolnwood, IL, March 1988.
6. Johnson, Colin C. *Solder Paste Technology—Principles and Applications.* Blue Ridge Summit, PA: TAB Books, 1989.
7. Keyser, Carl A. *Materials Science in Engineering,* Columbus, OH: Charles E. Merrill, 1974.
8. Manko, Howard H. *Soldering Handbook for Printed Circuits and Surface Mounting.* New York: Van Nostrand Reinhold, 1986.
9. ———*Solders and Soldering.* New York: McGraw-Hill, 1979.
10. Morency, Daniel. "A Discussion of SMT Solderability Issues and Relationships to Lead Finish." *Surface Mount Technology,* June 1991, pp. 30–34.
11. Prasad, Ray, P. *Surface Mount Technology—Principles and Practice.* New York: Van Nostrand Reinhold, 1989.
12. Wolverton, Mike. "Component Solderability." *Circuits Assembly,* March 1991, pp. 34–42.

6
Adhesive and Solder Paste Application Methods

GLOSSARY

Aperture   An opening in a screen or stencil.
Fixed Squeegee   A squeegee blade that is locked in a particular horizontal position.
Flexible Stencil   A stencil that is bonded to the frame with a wire mesh.
Floating Squeegee   A squeegee blade that is attached to a floating point, which allows the squeegee blade to follow the contour of the PCB surface during printing.
Off-Contact   A printing condition where the screen or stencil is positioned a certain distance from the PCB surface. The squeegee blade forces the screen or stencil into momentary contact with the PCB surface.
On-Contact   A printing condition where the stencil is positioned so it always contacts the PCB surface.
Screen   A screen mesh that is covered with an emulsion. The emulsion has openings that match the land patterns on the PCB surface. During printing, the adhesive or solder paste is forced through these openings onto the PCB.
Snap Off   The distance from the bottom of the screen or stencil to the top of the PCB. Used with off-contact printing.
Squeegee Blade   A plastic or metal blade used to pull the adhesive or solder paste over the screen or stencil.
Stencil   Thin sheets of brass or stainless steel with chemically etched openings that match the land patterns on the PCB surface. During printing the adhesive or solder paste is forced through these openings onto the PCB.

6.0 INTRODUCTION

Two methods are commonly used to apply solder paste and adhesive to a printed circuit board (PCB): printing and dispensing. Each method is an

TABLE 6.1 Printing Versus Dispensing

Method	Attributes	Concerns
Printing	Fast process	Flat surfaces only
	Soft tooling (printer)	Hard tooling (stencil)
	Good volume control	Open system
	Low maintenance (printer)	High maintenance (stencil)
		Slow set-up
Dispensing	Soft tooling (machine)	Slow process
	Irregular surfaces	Hard tooling (program)
	Good volume control	High maintenance (syringe)
	Closed system	Dot size limitations
	Low maintenance (machine)	
	Fast set-up	

acceptable approach to material application. Which one is used depends on the requirements of the user. Table 6.1 reviews the attributes and concerns associated with printing and dispensing. Soft tooling is flexible; hard tooling is not flexible.

Printing and dispensing are all too often neglected, which is unfortunate, because they are probably responsible for more soldering defects than any other surface mount process. More attention has usually been directed toward the placement and reflow processes. However, printing and dispensing have more variables than any other surface mount process. If not accomplished correctly, these processes will defeat the remaining surface mount processes even if they are done properly. This chapter will review screen and stencil fabrication, squeegee blades, printing systems, printing methods and process parameters, printing defects, dispensing equipment, and dispensing defects.

6.1 PRINTING

Printing is a process driven by fluid dynamics. It can repeatedly apply controlled quantities of material to the surface of a PCB. In general, the printing process is fairly simple. The PCB surface is placed near (off-contact) or in contact (on-contact) with the bottom surface of a screen or stencil. Material is pulled across the surface of a screen or stencil by a squeegee blade. Apertures in the screen or stencil fill with material. The material also makes contact with and adheres to the PCB surface. Finally, the screen or stencil is lifted away from the PCB, leaving the printed image on the PCB surface.

The name "screen printing" is left over from the days when only screens were used to apply inks, adhesives, and solder pastes. A stencil is now used for most solder paste applications. The printing process is usually associated with solder paste; however, it has been used successfully to apply surface mount adhesives. The advent of fine-pitch technology has added another level of complexity. Today, the printing process and solder paste technology are being pushed to their limits.

6.2 SCREEN AND STENCIL FABRICATION

6.2.1 Frames

Frames, which are made of aluminum, are either cast or extruded. Generally, the size of the frame dictates which method is used. Casting works well for smaller frames, but as the frame gets larger it becomes more difficult to cast. Larger frames are usually made using extruded aluminum that is welded together. Cast frames are slightly more accurate because they are machined after casting. Extruded frames are more flexible, and as a result they will conform to the printer clamping mechanism.

A screen is mounted to the frame by gluing the mesh directly to the bottom of the frame. Stencils are attached to the frame directly or by using a polyester or stainless steel mesh. When the direct method is used, the metal foil is glued to the bottom of the frame. When mesh is used, it is glued to the top of the metal foil and then to the bottom of the frame. The mesh attachment method, referred to as a flexible stencil, is favored because it pulls the metal foil tight and eliminates warp. Frame mounting guidelines are provided in Figure 6.1 (screens) and Figure 6.2 (flexible stencils).

6.2.2. Screens

A screen, as shown in Figure 6.3, is made by applying a photographic emulsion to a stainless steel wire mesh. The apertures are formed by curing most of the emulsion using high-intensity ultraviolet (UV) light. The emulsion that was covered by the master artwork does not cure and is washed away, forming the apertures.

The wire mesh is available in different sizes relating to the number of wires per inch and the diameter of the wire. An 80-mesh (80 wires per inch) screen with a wire diameter of 0.94mm (0.0037") is probably the most common. Other common mesh sizes are 60, 120, 150, 180, and 200. Wire diameters vary from approximately 0.05mm (0.002") up to 0.127mm (0.005"). The approximate wet print thickness (WT) can be determined by:

$$WT = (MT + ET) \times (AO \times PO)$$

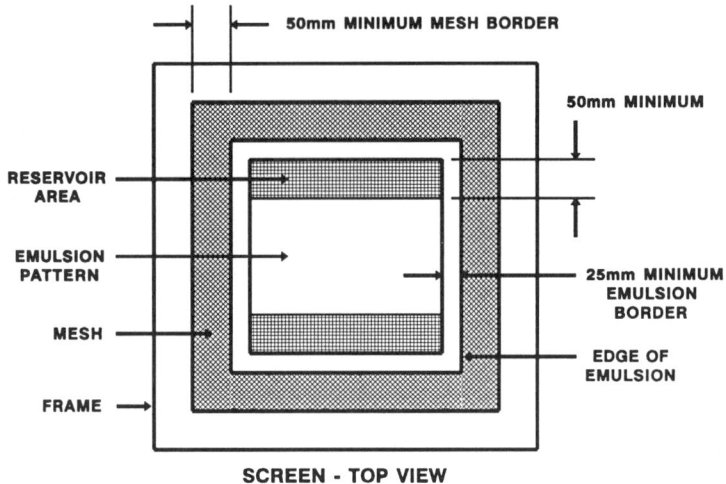

FIGURE 6.1. Example of a Properly Mounted Screen.

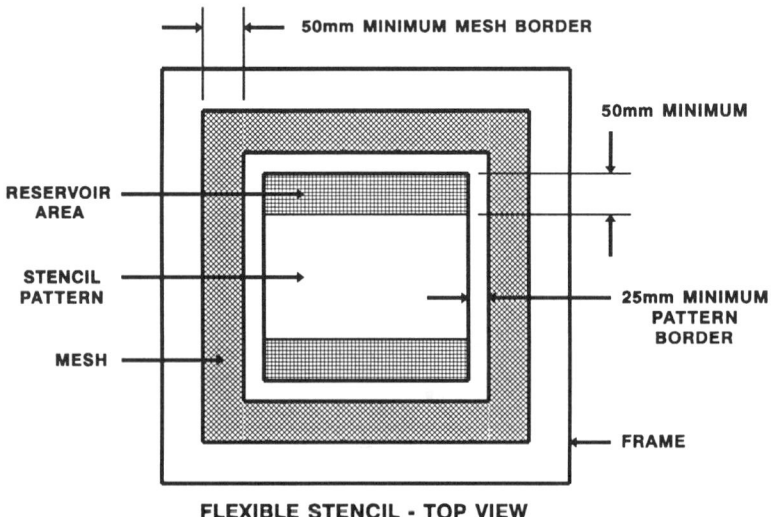

FIGURE 6.2. Example of a Properly Mounted Stencil.

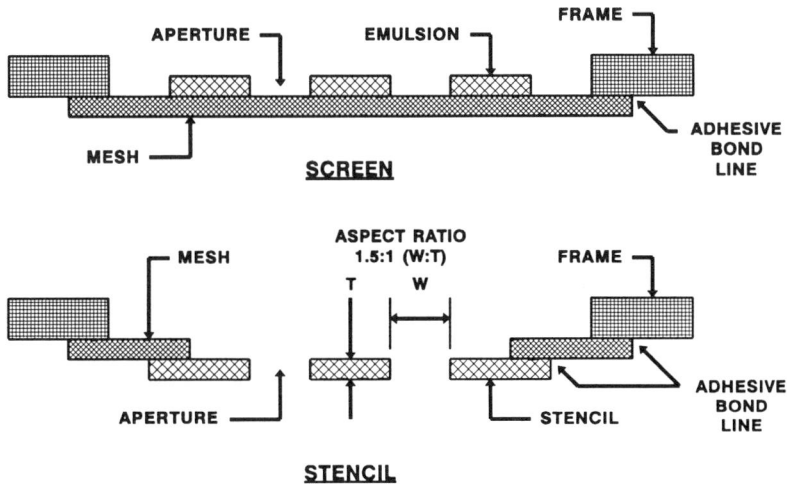

FIGURE 6.3. Section View Comparing Screens and Stencils.

where MT is the mesh thickness (2 × wire diameter), ET is the emulsion thickness, AO is the aperture open area and PO is the percent open area for a particular mesh size/wire diameter. Information about wire diameter, emulsion thickness, and open area can be obtained from the screen supplier.

The emulsion is applied to the wire mesh by a direct or indirect method. The direct method applies a liquid emulsion to the wire mesh, while the indirect method uses a solid preactivated film. In either case, the emulsion is cured, then activated, then cured again, and then washed to remove uncured emulsion.

For the most part, the use of screens has diminished in favor of stencils, at least for solder paste applications. For solder paste printing, screens are used mainly for prototype and low-volume applications where fine-pitch technology is not involved. Screens are still successfully used to print adhesives. Aperture design details can be found in sections 6.5.3 (adhesive) and 6.6.2 (solder paste).

6.2.3 Stencils

A stencil, as shown in Figure 6.3, is made by applying a liquid or dry film laminate mask to a metal plate. The mask covers the entire metal plate, except for those areas that are to be chemically etched away to form the apertures. After etching, the mask is removed. The metal plate is then attached directly to the frame for an all-metal stencil, or to a mesh that is attached to the frame

88   Applied Surface Mount Assembly

for a flexible metal stencil. The mesh attachment method is preferred because the mesh keeps the metal plate under tension, which prevents warp and twist.

Stencils are usually made from brass or stainless steel. Brass was the early favorite because it was easy to obtain and easy to etch. However, brass is quite soft with a relatively low tensile strength of about 27,000 lb. Thus, a brass stencil can be easily damaged during the printing process or by careless handling or storage. Stainless steel, on the other hand, has a high tensile strength of about 163,000 lb. Stainless steel stencils are more durable and resistant to damage. The main drawback of stainless steel is that it is more difficult to etch. See Figure 6.4 for an example of an etched side wall.

The aperture in the metal plate will increase in size by approximately 50% over the pattern on the artwork. This increase is a product of the chemical etching process and is called the etch factor. For example, a metal plate 0.254mm (0.010″) thick will have an increase of approximately 0.127mm (0.005″) in both the X and Y direction. The etch factor is controlled by the user or supplier modifying the artwork. Check with the stencil supplier to determine the best method to use to control the etch factor. Another import-

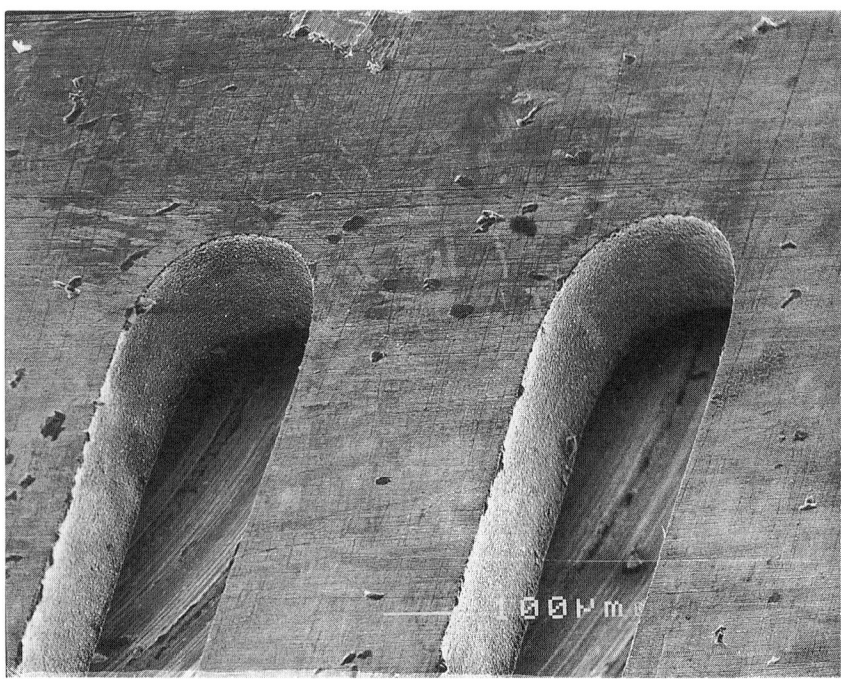

FIGURE 6.4.   Etched Stencil Side Wall.

ant factor is the aspect ratio, which is the ratio of the stencil thickness (T) to the narrowest width (W) of the aperture, as shown in Figure 6.3. The preferred minimum aspect ratio is 1 (T) to 1.5 (W). Lower aspect ratios can be achieved with more difficulty, but printing also becomes very difficult below this aspect ratio because the solder paste has difficulty releasing from the stencil. Etching can be done from one (single-sided) or both sides (double-sided). Single-sided etching produces a dome-shaped opening, where the opening on one side is larger than the opening on the other side. Double-sided etching produces an opening that is more uniform on both sides, which is preferred. One theory proposed that the dome shape, when placed so that the larger opening faced the PCB, allowed the solder paste to release from the stencil easier. This theory has been proven incorrect because the solder paste tends to fill the entire opening, which actually brings more surface area into contact with the solder paste, making it more difficult to release the solder paste. Aperture design details can be found in sections 6.5.3 (adhesive) and 6.6.2 (solder paste).

6.2.4 Alignment Targets

If the printer is not equipped with a vision system, it can be difficult to align the screen or stencil to the PCB. Manual alignment targets can be used as a set-up aid. The target can be almost any shape; a square approximately 2.54mm (0.100″) works well. Two targets are placed on the screen or stencil and two on the PCB, preferably in opposite corners. To use, simply load the PCB into the printer and align the screen or stencil targets to the PCB targets. The alignment targets can also be used as a process-control tool: The solder paste or adhesive applied to the targets can be inspected for alignment, thickness, and definition rather than inspecting the entire PCB.

6.3 SQUEEGEE BLADES

The squeegee blade is a simple yet critical printing tool. The forward movement and angle of the squeegee blade generates a force that pulls the material across the surface of a screen or stencil. The tip of the squeegee blade applies a shearing force that fills the screen or stencil apertures with material. When off-contact printing is used the squeegee blade also forces the screen or stencil into contact with the PCB surface. There are two types of squeegee blades: plastic and metal. Plastic squeegee blades are by far the most common. The metal squeegee blade, which may have some advantages over a plastic squeegee blade in certain applications, is a recent development. Most of the information in this section applies to plastic squeegee blades, since metal squeegee blades are new and there is a limited amount of experience with them.

The condition of the squeegee blade tip is critical. The best printing is achieved with a sharp, flat edge. During printing the tip will wear and develop a radius rather than a point. As the tip wears and the radius increases, more pressure will be required to obtain a clean wipe of the screen or stencil surface. When too much pressure is applied the tip may deform, decreasing the effective printing angle. The amount of wear will vary with the hardness and/or type of squeegee material and the printing surface with which the tip is in contact.

Squeegee speed for typical, non–fine-pitch applications ranges from 50 to 100mm/second (2 to 4"/second). Slower speeds are required for fine-pitch applications, usually in the 13 to 50mm/second (0.5 to 2"/second) range. The slower speed allows the solder paste to fill the small apertures properly. Squeegee pressure will typically range from 5 to 30 pounds, depending on the squeegee blade type and material, the off-contact distance, whether a screen or stencil is used, and the solder paste rheology.

The relationship between squeegee pressure and squeegee speed is important. In theory, there is a squeegee pressure and speed that will result in the optimum edge pressure and shearing action for each solder paste formulation. These two parameters are inversely proportional to each other. At any squeegee pressure, an increase in squeegee speed will decrease the pressure on the squeegee edge. Inversely, a decrease in squeegee speed will increase the pressure on the squeegee edge.

During the printing process there is friction between the squeegee blade and the screen or stencil. To help decrease the friction, apply adhesive or solder paste along the entire blade width. This will lubricate the blade edge and decrease friction. Also, the squeegee blade width should be approximately 50mm (2") wider than the PCB (25mm (1") per side). This keeps the amount of force applied to the stencil to a minimum, which prevents the stencil mesh from stretching.

Squeegee blades do not maintain their integrity forever. They should be replaced on a regular schedule as recommended by the supplier. When not in use they should be stored properly.

6.3.1 Plastic Squeegee Blades

Plastic squeegee blades are made from synthetic materials, such as polyurethane, that are resistant to the solvents used during the printing process. There are two common plastic squeegee blade shapes, diamond and rectangular, as illustrated by Figure 6.5. The diamond-shaped blade can be used to print in two directions (bidirectional), while the rectangular blade can only print in

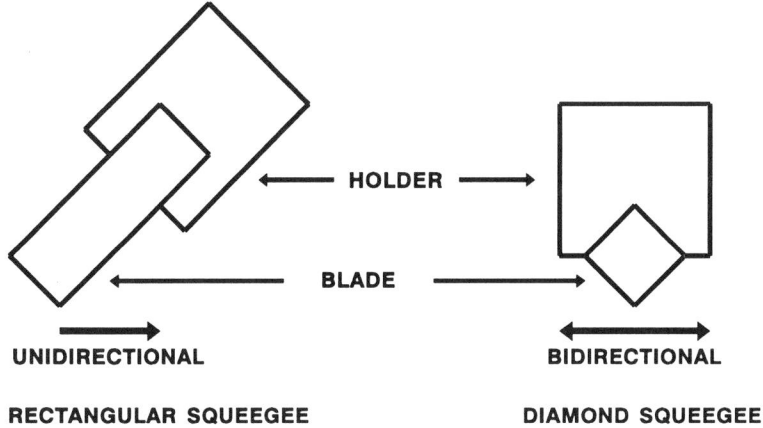

FIGURE 6.5. Common Squeegee Blade Shapes.

one direction (unidirectional). Diamond-shaped blades are fixed at a print angle of 45°. Rectangular-shaped blades are usually mounted in adjustable holders that allow the print angle to be varied. A 30° setting is common.

The hardness of the squeegee blade is measured in durometer on the shore A scale. For printing, squeegee blade material is available in a hardness range from 70 to 110 durometer. The hardness of the squeegee blade will vary with the application. A 70 or 80 durometer squeegee blade may be used with a screen and a low-viscosity material (850,000 CPS or less). A 70 durometer squeegee is not recommended for use with a stencil because it will conform to the surface of the stencil and scoop the solder paste out of the apertures. An 80 or 90 durometer squeegee blade is recommended for use with a stencil and a higher viscosity material (above 850,000 CPS).

As noted before, the condition of the squeegee blade edge is critical to successful printing. A sharp, smooth, uniform edge is desired. Plastic squeegee blades will wear considerably during use. The edge of the squeegee blade should be inspected daily, and replaced or resharpened when necessary. Remember, there are four usable edges on each squeegee blade. To keep track of which edges have been used, cut a V-shaped notch into an edge when it is no longer useable.

6.3.2 Metal Squeegee Blades

Metal squeegee blades are made from thin sheets of stainless steel or brass. They are intended primarily for use with stencils. At this time they are only available on a retrofit basis. Metal squeegee blades can only print in one direction (unidirectional). The printing angle varies from 30° to 45°.

Because metal squeegee blades remain flat during printing, they will not scoop solder paste out of the stencil apertures. Metal squeegee blades also offer the advantage of increased wear resistance. The edge of the blade will remain sharp, smooth, and uniform for an extended period of time. The blade edge should still be inspected at least weekly. The edge of the blade is usually impregnated with a lubricating material, which will decrease friction between the stencil and the blade. Compared to plastic squeegee blades, there is less friction between a metal squeegee blade and a stencil. This decreases stencil movement during printing, which decreases registration errors and increases stencil life.

6.4 PRINTING SYSTEMS

6.4.1 Print Modes

The various squeegee blade functions are described by three print modes: flood/print, print/flood, and print/print. The flood/print and print/flood modes are primarily used in conjunction with screens for printing low-viscosity materials, such as adhesives. The print/print mode can be used with screens or stencils for printing medium- to high-viscosity materials, such as solder paste. The flood/print and print/flood modes are used with the screen in the off-contact condition, while the print/print mode can be used with the screen or stencil in either the off-contact or on-contact condition.

The flood/print and print/flood modes use one unidirectional squeegee blade and one flood blade. Flooding is used to spread material over the surface of the screen, which begins the process of filling the apertures. The metal flood blade is positioned 0.1mm to 0.4mm (0.004" to 0.016") above the screen, directly across from the squeegee blade. Flooding is often required to help fill screen apertures because they are partially closed by the wire mesh, making it difficult to transfer material through the aperture. Once the flood phase is complete the squeegee blade presses down on the screen and moves across its surface, transferring the material through the aperture onto the PCB surface. The print/flood function acts in reverse of the flood/print function. The squeegee blade prints the material, and then the flood blade returns the material to the start position.

The print/print mode uses one bidirectional or two unidirectional squeegee blades. This mode can be used with the screen or stencil in the off-contact or on-contact position. The squeegee blade simply makes contact with the screen or stencil surface and pulls the material across it, transferring the material through the apertures onto the PCB surface. Contact printing, used primarily with stencils, places the stencil in contact with the PCB, as shown in Figure 6.6.

6.4.2 Contact and Off-contact Printing

Off-contact (snap off) printing, as shown in Figure 6.6, is used more often with screens than stencils; in fact, it is mandatory when a screen is used. A Snap-off setting is used with screens to prevent smearing that would occur during the flood mode. Once the screen is flooded, the squeegee blade forces the screen directly below the blade into momentary contact with the PCB, where a shearing action transfers the material to the PCB surface. Off-contact printing can also be used with stencils that are attached to the frame with a wire mesh, although it is more common to use an on-contact print. A Snap-off setting is not usually required when printing standard-pitch surface mount components; however, there may be some benefit when printing fine-pitch components. Off-contact printing may aid the release of the solder paste from the fine-pitch openings. An alternate approach is the use of "negative" snap off, which is accomplished by forcing the stencil into contact with the PCB so that it bows upward. This method is difficult to apply and control on most printers, and it may cause excessive stencil wear or even damage the stencil.

In most cases, on frames up to approximately 380mm (15") square, the snap-off distance can typically be set at 0.5mm to 1.25mm (0.020" to 0.050"). When using larger frames the snap-off distance may have to be increased up to approximately 2.5mm (0.100") to compensate for the increased surface area of the screen or stencil.

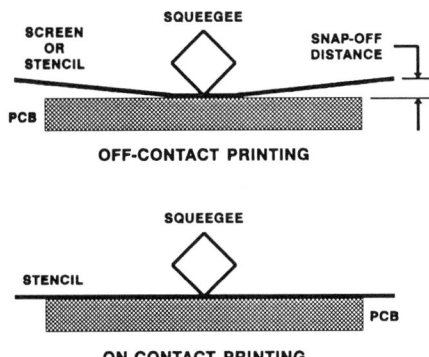

FIGURE 6.6. On-Contact versus Off-Contact Printing.

6.4.3 Print Head Assembly

The print head assembly consists of two separate assemblies: the frame support system and the PCB support system. Two fundamental concepts are used for frame support systems: the clamshell and the vertical column, as displayed in Figure 6.7. The clamshell design is hinged at the rear of the printer. The frame support system pivots up to allow access to the print area and pivots down to allow printing. The vertical column design maintains the frame support system in a horizontal position, while the PCB support system moves the PCB into the print position. The clamshell system is less expensive, but the vertical column system is more accurate.

The frame support system registers and secures the screen or stencil frame during printing. In addition, the frame support system in a vision-based printer aligns the stencil to the PCB.

The PCB support system registers, secures, and supports the PCB during printing using mechanical clamps, a vacuum plate, or sometimes a combination of both. The vacuum plate has been the primary method for holding the PCB in position during printing. However, there are some major disadvantages to the vacuum plate concept. The vacuum holes must be located away from holes in the PCB to avoid loss of vacuum. If the

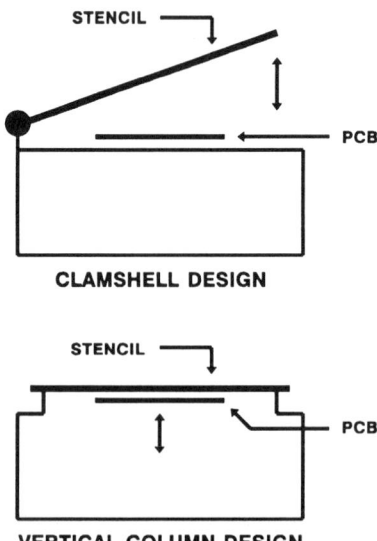

FIGURE 6.7. Clamshell versus Vertical Column Print Heads.

PCB has components on both sides, the vacuum plate for printing side two must be milled out to allow clearance for the components on side one. Both of these issues require the use of a custom vacuum plate for each PCB. Also, the vacuum plate alone may not be enough to hold a large PCB securely during printing; some mechanical assistance may be required. As a result of these problems some printers now use a mechanical clamping system in place of the vacuum plate. These designs consist of an edge-hold system, which clamps the PCB by its edges, and support pins, which can be placed at various locations under the PCB to provide adequate support during printing.

Positioning is accomplished with one of two methods: tooling pins or edge registration. Normally, positioning is accomplished by mating tooling holes in the PCB with tooling pins on the printer. Edge registration is accomplished by placing the edge of the PCB against a stop. However, this is a less repeatable method because the distance from the PCB edge to the PCB features (land patterns, pads, etc.) varies more than the distance from the PCB tooling holes to the PCB features.

6.4.4 Manual Printers

Manual printers, priced from approximately $2,000 to $10,000, are the simplest and least expensive systems available. They are used for very low production or research and development applications. The loading and unloading of the PCB is accomplished manually by the operator. The squeegee blade can be hand held or attached to the printer, with the print stroke being performed manually by the operator. Alignment of the stencil to the PCB is always performed manually. Positioning of the PCB is achieved with tooling pins or edge registration. These printers depend greatly on the skill of the operator, some to a greater extent than others. Even with the best operator the results will vary because many of the printing parameters vary each time a PCB is printed. A well-designed manual printer is displayed in Figure 6.8. Note the attached squeegee blade and the clamshell design. The operator pulls the squeegee handle to complete the print stroke.

6.4.5 Semiautomatic Printers

The semiautomatic printer, with a price range from approximately $15,000 to $50,000, is the most common printer in use today. These versatile printers are used primarily in low- to medium-volume manufacturing. They are actually very similar to manual printers. The PCB is still loaded and unloaded manually. The primary improvement deals with the print head. These systems

96   Applied Surface Mount Assembly

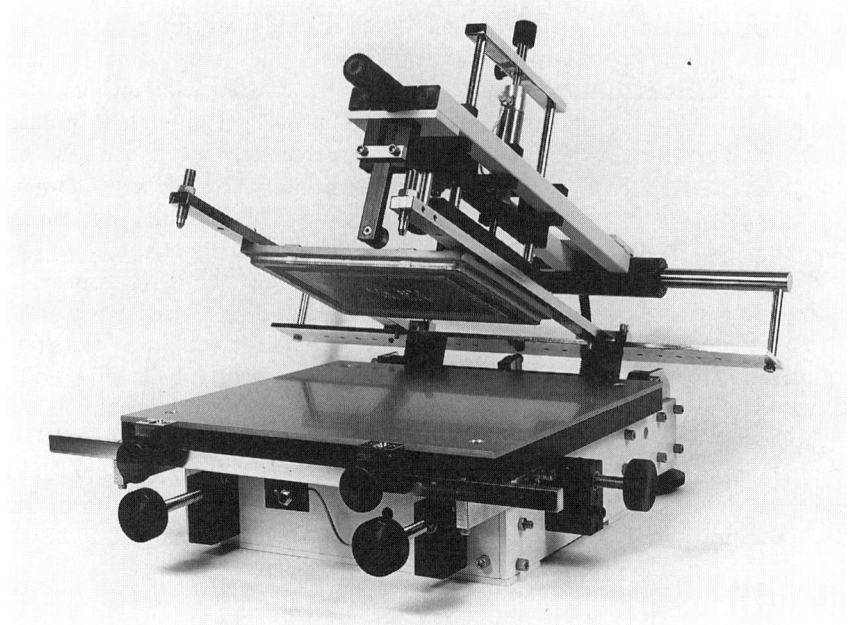

FIGURE 6.8.   Example of a Manual Printer.

have better control over parameters such as print speed, squeegee pressure, squeegee angle, print stroke length, and snap-off distance. Tooling pins or edge registration are used to position the PCB. Stencil to PCB alignment is usually accomplished manually. More sophisticated systems use manual vision alignment to help the operator align the stencil to the PCB. The semiautomatic printer shown in Figure 6.9 is a good example of a printer with a clamshell design.

6.4.6 Automatic Printers

Automatic printers, priced from $75,000 to $250,000, perform most or all functions automatically. They are intended for medium- to high-volume manufacturing or for applications requiring a great deal of accuracy and repeatability, such as fine-pitch printing. These printers are all microprocessor controlled. The PCB is loaded and unloaded automatically using an edge-hold conveyor system. Parameters such as squeegee speed, squeegee pressure, print stroke length, and snap-off distance are programmable. The PCB is positioned using tooling pins or edge registration. Some printers have

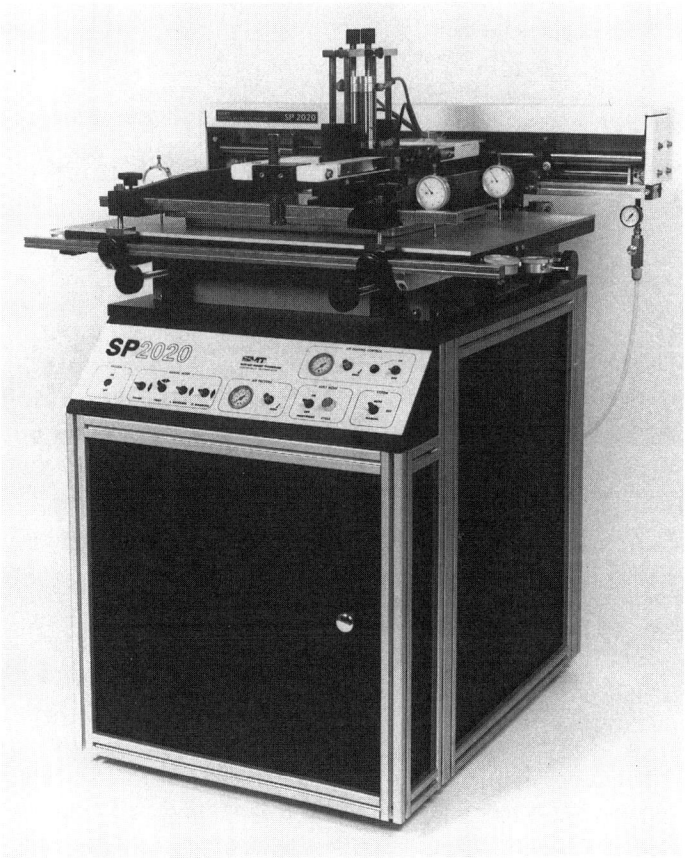

FIGURE 6.9.   Example of a Semi-Automatic Dispenser.

the ability to automatically align the PCB and stencil using a vision system. When a vision alignment system is used, edge registration is popular because of ease of set-up. The vision system will neutralize the increased error associated with edge registration. Stencil to PCB alignment is performed manually or with the aid of a vision alignment system. Fine-pitch technology has mandated the use of automatic stencil/PCB vision alignment systems. Refer to section 6.13 for information on vision alignment systems. Figure 6.10 shows an automatic printer with vision alignment. Notice the monitor located at the left rear of the printer.

FIGURE 6.10. Example of a Automatic Vision Printer.

6.5 PRINTING ADHESIVES

Surface mount adhesives are used to attach certain surface mount components to the secondary (solder) side of a PCB until they can be wave soldered. Two methods are commonly used to apply surface mount adhesives to a PCB: dispensing and screen printing. Another method, pin transfer, has been used in limited applications. The most common and flexible method by far is dispensing (which is discussed in section 6.10), but under certain conditions screen printing can be an effective approach.

6.5.1 Component Range

The screen printing method is best suited to rectangular chip components in the size range of 1206 (0.120″ × 0.060″) or larger. Small outline transistors (SOTS) require a small dot size and a standoff that is difficult to achieve with this process. Integrated circuits, such as the SOIC and PLCC, also have a standoff that is difficult to achieve with the printing process. Attachment of these components can be accomplished better with a pressure syringe, which can apply the required adhesive volume and thickness.

6.5.2 Adhesive Application

A print thickness of 0.2mm (0.008") is adequate to securely hold the component. Below 0.15mm (0.006") the print thickness is too thin; there is not enough adhesive to hold the component in place. A print thickness above 0.25mm (0.010") does not hold the component any better than the 0.2mm (0.008") print thickness, and after component placement the thicker adhesive deposit may slump onto terminations, leads, and lands.

For maximum strength, the adhesive deposit should be under the component and extend past the component width at least 0.5mm (0.020") on both sides, which will allow the adhesive to bond to the sides of the component as well as the bottom. The adhesive deposit should be at least 0.5mm (0.020") away from any land. This will ensure that after component placement the adhesive will be approximately 0.25mm (0.010") away from any land. This is important because the adhesive will tend to slump and spread during the curing process. It is helpful to use a UV cure adhesive to reduce slump and spread. The UV cure quickly forms a hard cover over the adhesive, which will reduce its ability to slump and spread, then curing can be completed with heat.

Screens or stencils can be used to apply the adhesive. The best results will probably be achieved with the stencil. Only a stainless steel stencil should be used. An interaction with a brass stencil may cause some adhesives to cure on the stencil.

6.5.3 Screen/Stencil Aperture Design

The screen or stencil thickness should be 0.006" minimum and 0.010" maximum (keep in mind the 1.5 to 1 aspect ratio). The aperture size can be determined using the following formulas. See Figure 6.11 for an example of the aperture orientation.

$$\text{LENGTH} = W + 0.040''$$
$$\text{WIDTH} = S - 0.040''$$

where

$W =$ Width of component
$S =$ Span (distance between lands)

Example:

Determine aperture size for a 1206 component.

$$\text{LENGTH} = 0.060'' + 0.040'' = 0.100''$$
$$\text{WIDTH} = 0.070'' - 0.040'' = 0.030''$$

100   Applied Surface Mount Assembly

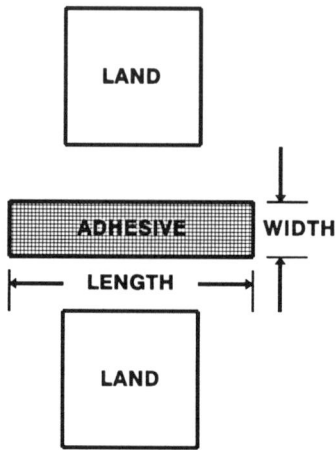

FIGURE 6.11.   Adhesive Aperture Orientation.

6.6 PRINTING SOLDER PASTE

6.6.1 Solder Paste Application

The objective of solder paste printing is to apply solder, in paste form, to the lands on the PCB. After the placement process the solder paste will hold the components in place until reflow occurs, at which time the mechanical and electrical interface between the component and the PCB is formed. Solder paste volume is controlled by the thickness of the screen or stencil and the size of the aperture. The aperture is often modified (decreased in size) to change the volume of solder paste applied to the lands. Stencils are preferred over screens because they achieve better results and are more durable, which justifies their added cost.

6.6.2 Screen/Stencil Design

Solder paste volume is directly related to screen/stencil thickness and aperture size, and unfortunately there are no industry standard recommendations relating to these areas. Testing is being done within the industry to establish the relationship between solder paste volume and solder joint reliability, however this information is not yet available. In the interim, some general guidelines can be used that will achieve successful results. Gull-wing leads require the least amount of solder; rectangular and cylindrical terminations require a moderate amount of solder; and J leads require the most solder. There is certainly a dilemma when it is necessary to apply solder paste for all of the lead configurations, especially when fine-pitch leads are involved.

Solder paste printing has become a compromise. There are many different lead configurations on a PCA, each requiring a different solder paste volume. It is becoming increasingly difficult to meet the needs of each lead type. A print thickness of 0.2mm (0.008″) is the most common and is adequate for most applications. A 0.25mm (0.010″) print thickness is preferred by some when J leads are involved (1 to 1 aperture). Fine pitch works best with a 0.15mm (0.006″) print thickness (1 to 1 aperture); however this may cause problems with other lead types, particularly J leads. The aperture size can be altered to achieve a specific volume. Table 6.2 is intended as a starting point for aperture design. These guidelines may be refined further depending on the application. Please review the results carefully.

To use Table 6.2 multiply the length and width of the land by the aperture size noted in the table.

Example:

Determine the aperture size for a gull-wing lead that has a land length of 0.060″ and a width of 0.025″ when using a 0.010″ material (stencil) thickness.

$$0.060″ \times 0.8 \, (= 0.048″)$$
$$0.025″ \times 0.8 \, (= 0.020″)$$

The aperture size would be 0.048″ by 0.020″.

One aperture that is different is the aperture for the J lead when the 0.006″ material thickness is used. With a 0.006″ print, J leads do not really receive

TABLE 6.2 Aperture Size Guidelines

Lead Type	Material Thickness	Land Area	Aperture Size
Rectangular	0.2mm (0.008″)	1.0	1.0
	0.25mm (0.010″)	1.0	0.8
	0.15mm (0.006″)	1.0	1.0
Cylindrical	0.2mm (0.008″)	1.0	1.0
	0.25mm (0.010″)	1.0	0.8
	0.15mm (0.006″)	1.0	1.0
Gull Wing	0.2mm (0.008″)	1.0	1.0
	0.25mm (0.010″)	1.0	0.8
	0.15mm (0.006″)	1.0	1.0
Gull Wing (fine pitch)	0.2mm (0.008″)	1.0	0.7
	0.25mm (0.010″)	1.0	N/A
	0.15mm (0.006″)	1.0	0.9
J Lead	0.2mm (0.008″)	1.0	1.0
	0.25mm (0.010″)	1.0	1.0
	0.15mm (0.006″)	1.0	1.2

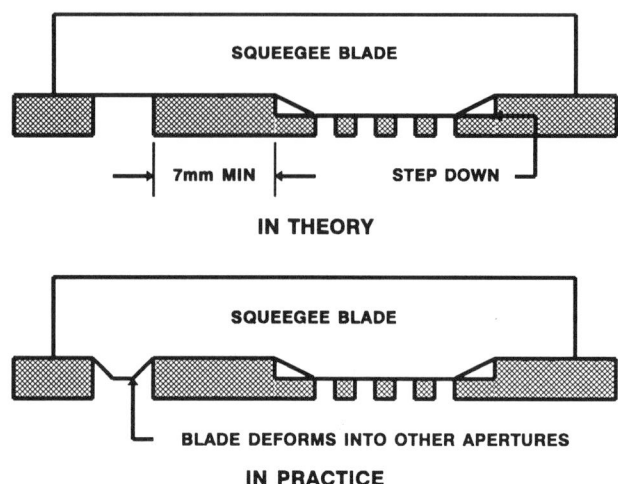

FIGURE 6.12. Bi-Level Stencil: Theory versus Practice.

enough solder so it is necessary to make the aperture oversized. Also, a 0.010″ stencil cannot successfully print fine pitch because it violates the 1.5 to 1 aspect ratio discussed earlier.

Another method for controlling the solder paste volume is the bi-level or step stencil, as shown in Figure 6.12. This is a stencil that has selected areas etched thinner than the rest of the stencil. This process is primarily used for fine-pitch applications. A typical application would be to use a 0.2mm (0.008″) thick stencil and etch the fine-pitch areas to a 0.15mm (0.006″) thickness. In theory, a soft (70 to 80 durometer) squeegee blade is used so it will deform into the step-down and shear the solder paste off at the 0.15mm (0.006″) thickness. In practice, however, the squeegee blade also deforms into other apertures and scoops the solder paste out, causing insufficient solder paste application in those areas. A distance of at least 6.35mm (0.250″) is required between the fine-pitch apertures and any other apertures.

6.7 PROCESS PARAMETERS

6.7.1 Equipment Parameters

Optimum parameters may vary between printers.

- Off-contact mode for screens
- Contact mode for stencils
- Plastic squeegee blade hardness, adhesive: 70 to 80 durometer

- Plastic squeegee blade hardness, solder paste: 80 to 90 durometer
- Consider using a metal squeegee blade
- Print speed, standard: 2 to 4"/second
- Print speed, fine pitch: 1 to 2"/second
- Squeegee pressure: 5 to 25 pounds depending on printer type
- Floating squeegee head preferred

6.7.2 Process Control

Process control is very important. The following items should be monitored on a continuous basis.

- Printer performance
- Printer settings
- Temperature of solder paste or adhesive
- Humidity near printer
- Print thickness
- Print registration and definition

6.8 PRINTED CIRCUIT BOARD DESIGN

The selection and application of solder mask (see Chapter 3) is very important to the printing process. Ideally, the best result would be solder mask that was perfectly flat and approximately 0.025mm (0.001") above the surface of the lands. A flat surface is especially desirable when a stencil is used, so a seal will form between the PCB and the stencil. This will decrease squeeze out and improve print definition.

There are three types of solder mask to select from: conventional wet film, photoimageable dry film, and photoimageable wet film. Most of the conventional wet film processes do not have the registration control required for surface mount PCB's, and they are usually not a good choice for this application. The photoimageable processes will provide the best registration, flatness, and thickness control. The photoimageable dry film process tends to be a little thicker than desired, usually 0.08mm to 0.1mm (0.003" to 0.004") above the lands, but it will supply a very flat surface. Thinner dry film material is becoming available. The photoimageable wet process will provide a thin application, usually 0.025mm to 0.05mm (0.001" to 0.002") above the lands. Depending on which material is used, it will also supply a surface finish that ranges from moderately flat to very flat. Circuitry routed between the lands will also affect the surface flatness. Conductors routed between lands should be balanced so a "teeter-totter" effect is not created, with the conductor acting as a pivot point; or use a solder mask that can conform to the conductor and provide a flat surface.

Do not apply silk screen ink (legend) to areas that will have adhesive applied to them. The silk screen ink creates an irregular surface, which will affect print thickness and the seal between the stencil and PCB.

6.9 PRINTING DEFECTS

Printing defects can be defined by six categories, as indicated below and in Figure 6.13.

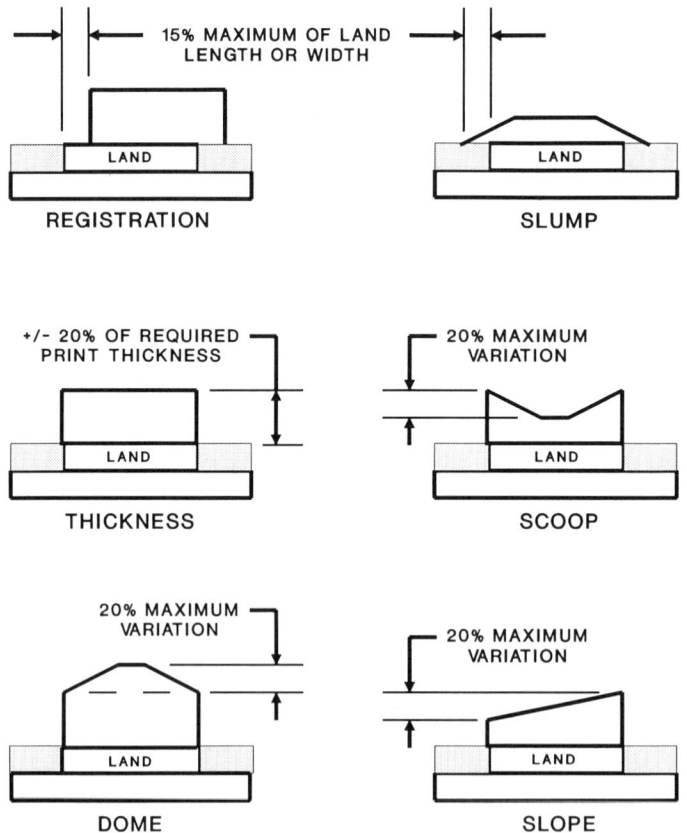

FIGURE 6.13. Common Printing Defects.

Registration

This involves the alignment of the screen or stencil to the area where material is to be deposited, either the land (solder paste) or the span between the lands (adhesive). The maximum allowable registration error should be 15% of the land width or length for solder paste application, and 15% of the aperture width or length for adhesive application.

Slump

This is a material-related defect. Slump is caused by an adhesive or solder paste viscosity that is too low for the application, or exposure to excessive heat, which alters the solder paste or adhesive viscosity. The result of this defect is solder bridging between leads, especially with fine-pitch components, or contamination of the solder joint by the adhesive. The amount of slump should be limited to 15% of the land length or width for solder paste, and 15% of the aperture length or width for adhesives.

Thickness

The final print thickness should not vary more than ±20% of the desired print thickness. This defect can be affected by three other defects: scoop, dome, and slope. The result of a thin application of material can be insufficient solder in the solder joint or an insufficient amount of adhesive to hold the component in place. An application that is too thick can result in solder joints with excessive solder or bridging between leads; and an excessive amount of adhesive can result in contaminated solder joints.

Scoop

This is the result of excessive squeegee pressure, a squeegee blade that is too soft, and/or an aperture that is too large. This defect can cause a solder joint to have insufficient solder or an adhesive deposit to have insufficient adhesive to hold the component in place. The amount of scoop should be limited to a maximum variation of 20% from the high point to the low point.

Dome

This is a result of incorrect squeegee blade height adjustment and insufficient squeegee pressure. This will increase the material volume. This may cause bridging between leads, solder joints with excessive solder when solder paste is printed, or contaminated solder when adhesive is printed. The amount of variation should be limited to 20% of the print thickness.

Slope

This can result from excessive squeegee pressure and is also common in printing fine pitch when the long dimension (length) of the aperture is parallel to the squeegee blade. This defect is more common with solder paste. The result of this defect would be insufficient solder in the solder joint. The amount of variation should not exceed 20% from the high point to the low point.

6.10 DISPENSING

Dispensing is also a process driven by fluid dynamics. As with printing, it can repeatedly apply controlled quantities of material to the surface of a PCB. This is accomplished with equipment that ranges from hand-held syringes to automatic dispensing equipment. The most common application is adhesive dispensing, where the adhesive is used to hold the component in place until soldering is complete. In the past several years solder paste dispensing also has become an acceptable process, made possible by improvements in equipment and solder paste.

The material to be dispensed is usually purchased in a syringe, with the 10cc capacity being the most popular. Several methods are used to force the material from the syringe. Positioning is accomplished by hand or by an X-Y table or head. One major advantage of dispensing is the ability to vary dot height, which cannot be accomplished with a printing process.

6.11 DISPENSING METHODS AND EQUIPMENT

6.11.1 Syringes

A syringe assembly is made from three items: the barrel, the plunger, and the needle. The barrel is usually manufactured from high-density polyethylene in sizes ranging from 3cc to 30cc. Most barrels have a moderate taper, starting at the plunger end. The plungers are made from wax or rubber. The needles are fabricated from plastic or a combination of plastic and metal. Needles can be obtained with orifice diameters ranging from 0.15mm to 1.6mm (0.006" to 0.063"). The plastic needles are usually tapered and may do a better job of dispensing the material, depending on the orifice diameter. Needles should be handled with great care to prevent damage. The preferred needle length is approximately 6.35mm (0.250"). Needles should never be cut because this deforms the tip and causes burrs. Also, needles should never be bent, as this restricts the flow of the material.

When a syringe is filled with material it is absolutely critical that air pockets are not formed. These air pockets will result in voids during the dispensing process. The more advanced dispensing systems can use vision imaging to inspect for voids on the PCB.

6.11.2 Dispensing Methods

Four methods are used for dispensing: pulsed air, peristaltic (pinch tube) valve, rotary displacement pump, and piston displacement pump. The selection of the proper dispensing method is critical.

A pulsed-air system uses timed air pressure (approximately 40 PSI) to apply a force to the plunger in the barrel. This pressure forces the plunger to move, which forces a specific amount of material out of the needle. These systems, which are very common, are inexpensive and easy to maintain because there are no valves or pumps involved; however they are not as accurate and repeatable as other systems. They are suitable for dispensing adhesives. There is a concern when this method is used with solder paste. Under high pressure the metal powder separates from the vehicle, the powder is forced to the back of the tube, resulting in more vehicle and less powder being dispensed.

A peristaltic, or pinch tube, valve uses a syringe as a reservoir for the material to be dispensed. The material in the syringe is forced, using low pressure (5 to 15 PSI), to a piece of plastic tubing, which is initially pinched at both ends by two valves. The top valve is opened, allowing the material to flow into the tube. The top valve is then closed and the bottom valve is opened, at which time a piston pushes on the tube, forcing the material out of the needle. Since this valve uses low pressure, solder paste powder separation is not a major problem. However, problems relating to the plastic tube do occur. Pressure from the valve causes the tube to stretch and crack. The initial problem caused by the tube stretching is volume control, but the tube will eventually fail due to stress cracking.

A rotary displacement pump also uses a syringe as a material reservoir. The material is forced, under low pressure (5 to 15 PSI), to the entrance of a rotary screw. The controlled rotation of the rotary screw transfers the material from the entrance to the end of the needle. Rotary screw pumps can dispense different volumes of material using software commands, so different volumes can readily be applied on the same PCB. They can also dispense continuous beads of material. A concern occurs when using solder paste: the metal powder can separate from the vehicle when inside the rotary screw.

A piston displacement pump pushes the material from a chamber through the needle. A syringe is used as a reservoir for the material to be dispensed. The material is forced, under low pressure (5 to 15 PSI), through a passageway to the pumping chamber. When the piston strokes upward a negative pressure is created in the pumping chamber, allowing the material to flow into the pumping chamber. When the piston strokes downward the chamber inlet is closed off and the piston forces the small amount of material through the needle and ejects it onto the PCB. The piston displacement pump has very good volume control, but continuous beads cannot be formed. The piston displacement pump is, overall, the best dispensing method.

For proper dispensing the needle tip should be positioned one-half of the needle's inside diameter above the PCB surface. Several Z-axis sensing methods have been developed to achieve this positioning. Contact-type Z-axis sensors use a mechanical probe to detect the PCB surface and position the needle tip. Noncontract-type Z-axis sensors use a combination laser and vision system to determine the location of the PCB surface. The needle tip is then positioned accordingly. Programming allows different Z-axis heights to be used on the same PCB, which permits variation in dot size and volume.

Material temperature is very important because as the temperature increases the viscosity decreases. Most dispensing systems can provide temperature-control capability that can maintain the material at the proper temperature.

When dispensing adhesive to hold rectangular and cylindrical components in place, use a double dot, placing one dot on each side of the component. This will permit adhesive to flow up the side of the component, increasing the component surface area in contact with the adhesive and significantly increasing the bond strength.

6.11.3 Manual Dispensers

Manual dispensers range in price from approximately $200 to $1,000. They are used in prototype, rework, and low-volume applications that do not require great accuracy and repeatability. Manual dispensers rely heavily on the skill of the operator, since the operator is the positioning mechanism. These units use primarily the pulsed-air method to dispense material.

6.11.4 Semiautomatic Dispensers

Semiautomatic dispensers, as displayed in Figure 6.14, range in price from approximately $5,000 to $25,000. They are used for prototype and low-to-medium-volume applications. These systems, which use either an X-Y table or an X-Y head, can be small tabletop units or stand-alone machines. The PCB is loaded and unloaded manually. Tooling pins or edge registration are used to position the PCB. Vision alignment may also be used. See section 6.13 for information about this method. These systems use stepper motors, servo motors, and encoders to achieve good positional accuracy and repeatability. A vacuum plate or mechanical fixture may be used to keep the PCB flat

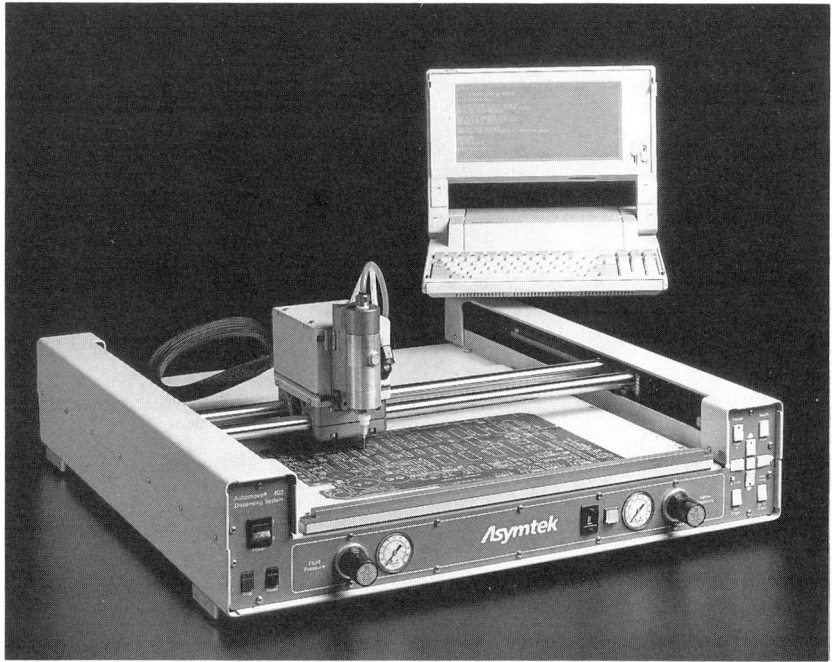

FIGURE 6.14. Example of a Semi-Automatic Dispenser.

during the dispensing operation. As a general guideline, the system accuracy should be equal to 10% of the smallest dot diameter. For example, if the dot diameter is 1.3mm (0.050″) then the positional accuracy should be 0.13mm (0.005″), but if the dot diameter is 0.635mm (0.025″) then the positional accuracy should be 0.0635mm (0.0025″). Any of the four dispensing methods are used.

6.11.5 Automatic Dispensers

Automatic dispensers, as represented in Figure 6.15, range in price from approximately $40,000 to $100,000. They are used primarily for medium- to high-volume applications. These stand-alone systems use an X-Y head positioned by stepper motors, servo motors, and encoders. Tooling pins are usually used to position the PCB. Vision alignment is also being used on these systems. See section 6.13 for information about this method. These systems are intended for use in automated manufacturing lines and they use a conveyor to transport the PCB in and out. A push plate, mechanical fixture,

110   Applied Surface Mount Assembly

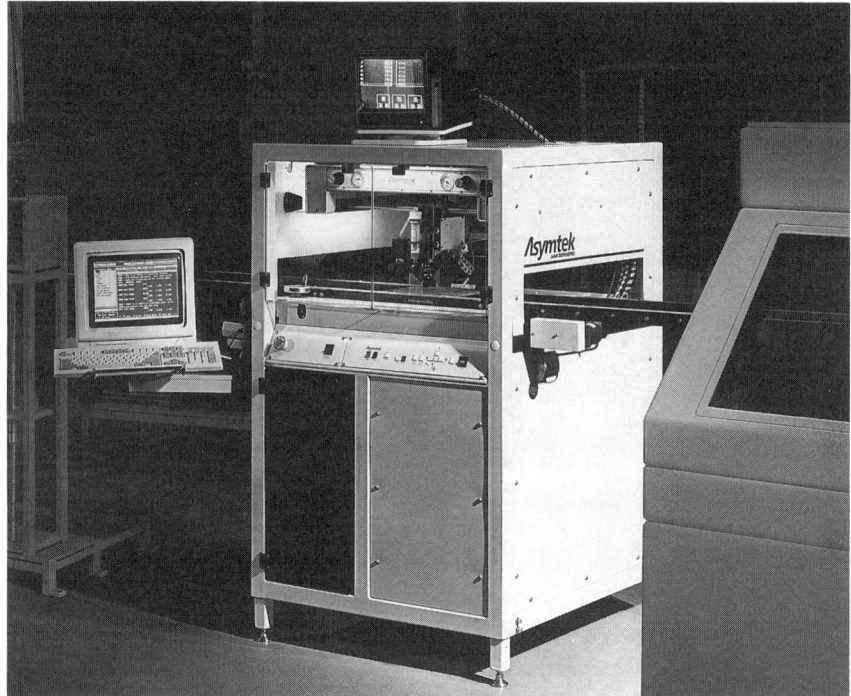

FIGURE 6.15.   Example of a Automatic Dispenser.

or vacuum plate is used to keep the PCB flat during the dispensing operation. As with semiautomatic systems, the accuracy should be equal to 10% of the smallest dot diameter. These systems can use any of the four dispensing methods.

6.12 DISPENSING DEFECTS

Dispensing defects can be defined by six categories, as indicated below and in Figure 6.16.

Registration
This involves the alignment of deposited material to the area where material should be deposited, either the land (solder paste) or the span between the lands (adhesive). The allowable registration error should be 25% of the dot diameter.

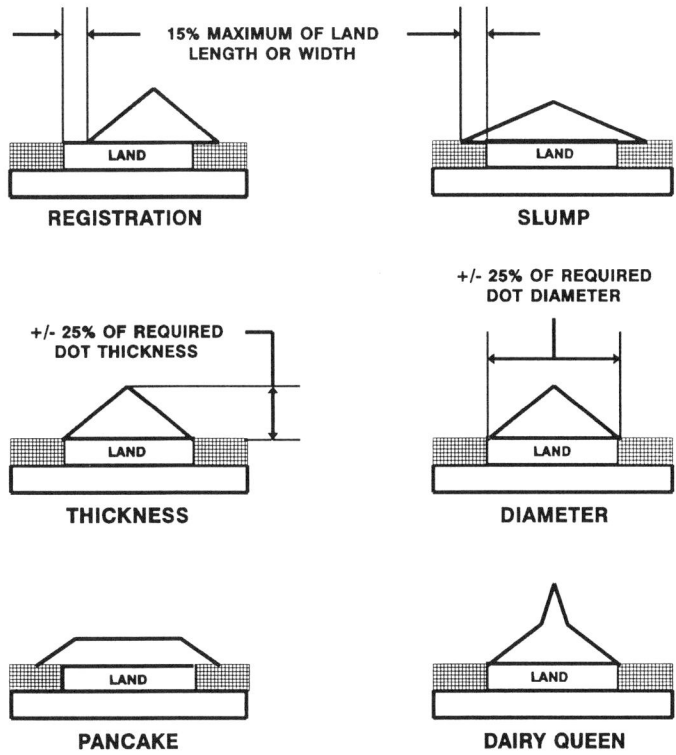

FIGURE 6.16. Common Dispensing Defects.

Slump
This is a material-related defect. Slump is caused by an adhesive or solder paste viscosity that is too low for the application, or exposure to excessive heat, which alters the solder paste or adhesive viscosity. The result of this defect is solder bridging between leads or contamination of the solder joint by the adhesive. Slump should be limited to 15% of the dot diameter.

Thickness
The final dot thickness should not vary more than ±25% of the desired dot thickness. The result of a thin application of material can be insufficient solder in the solder joint or an insufficient amount of adhesive to hold the component in place. An application that is too thick can result in solder joints with excessive solder or bridging between leads and an excessive amount of adhesive can result in contaminated solder joints.

Diameter

The final dot diameter should not vary more than ±25% of the required dot diameter. The result of an undersized dot is insufficient solder in the solder joint, or an insufficient amount of adhesive to hold the component in place. The result of an oversized dot is solder joints with excessive solder, or bridging between leads and adhesive-contaminated solder joints.

Pancake Effect

This occurs when the needle tip is too close to the PCB surface. The material is forced out to the side, resulting in a pancake deposit rather than a dot.

Dairy Queen Effect

This is an effect that only occurs with adhesives. It is the result of adhesive problems or incorrect dispenser parameters (for example, wrong needle diameter, improper distance between the needle tip and the PCB surface, and incorrect dispensing pressure).

6.13 VISION ALIGNMENT

Vision alignment on printer and dispensing systems is becoming an important method for neutralizing standard PCB fabrication errors. Two types of vision systems are used: binary and gray scale. Both methods use vision alignment in the same fundamental manner. The vision system uses pattern recognition to locate fiducial marks or other features on both the PCB and stencil. The vision system then determines the X, Y, and theta offsets between the PCB and the stencil. Using this information, the print head then aligns the stencil to the PCB. Binary imaging, which is the older of the two methods, locates a feature using the contrast between black and white images on the stencil or PCB. The system uses an algorithm to determine the centroid of a white feature in a binary image. Binary imaging has been popular because it requires only a moderate amount of computing capability; however, it is sensitive to contrast and lighting changes. Gray-scale imaging operates in a manner similar to binary imaging. The important difference is that a gray scale is used, rather than just black and white. A gray-scale image is normalized, which means it is not as sensitive to contrast and lighting changes as a binary image. The drawback to gray-scale imaging has been its high demand for computing power. But, as computers become more powerful, this problem is becoming less significant.

Many patterns can be used for a fiducial, but a circle is a very good choice because it is symmetrical about its axis and thus is not susceptible to rotational offset. Equipment suppliers usually provide recommendations for fiducial patterns. See Chapter 4 for examples.

REFERENCES

1. Babiarz, A. J. "Adhesive Dispensing for Surface Mount Assembly." *Printed Circuit Assembly*, July 1990, pp. 8–11.
2. Buttars, Scott. "Parameters for Solder Paste Printing." Proceedings, NEPCON West, February 1990, pp. 799–806.
3. Cavallaro, Kenneth. "Dispensing Adhesives for High Throughput SMT Assembly." *Electronic Packaging and Production*, May 1991, pp. 141–145.
4. ———and Marchitto, Michael. "Solder Paste Dispensing Versus Screen Printing." *Circuits Assembly*, October 1991, pp. 40–45.
5. Coleman, William E. "Design Parameters and Manufacturing Processes for Metal Mask Stencils." Proceedings, NEPCON West, February 1990, pp. 769–778.
6. Engel, Jack. "Advances in Automated Dispensing Equipment," *Circuits Manufacturing*, August 1990, pp. 51–54.
7. Enterkin, Robert. "Equipment/Material Synergism Key to Solder Paste Printing." *Electronic Packaging and Production*, May 1991, pp. 136–138.
8. Freeman, Gary. "Advances in SMT Screen Printing Equipment," Proceedings, NEPCON West, February 1990, pp. 779–790.
9. ———. "The Fine-Pitch Printing Process." *Circuits Manufacturing*, January 1990, pp. 50–55.
10. Jillings, Robert. "Screens for Solder Paste Printing," Journal of SMT, April 1991, pp. 11–21.
11. Johnson, Colin and Kevra, Joseph. Solder Paste Technology—Principles and Applications. Blue Ridge Summit, PA: TAB Books, 1989.
12. Lee, John, Jr. "Solder Paste Dispensing Materials and Requirements," *Electronic Packaging and Production*, November 1990, pp. 22–25.
13. Schatz, David and Freeman, Gary. "Automatic Alignment Aids Screen Printing," *Hybrid Circuit Techology*, February 1988.
14. Silver, William M. "Gray-Level Processing in Machine Vision," *ESD*, May 1987.

7
Component Placement

GLOSSARY

Actual Placement Rate The placement rate that the user will actually achieve in production.

CCD (Charged Coupled Device) camera A high-speed, high-density storage system that operates by the transfer of stored charges.

Encoder An electromechanical device used to translate the position of a drive mechanism.

Internal Centering A component centering mechanism that is located on the placement head.

External Centering A component centering mechanism that is located on the frame of the placement system.

Maximum Placement Rate The maximum rate of placement that can be achieved by a placement system. Usually based on minimum head and/or table travel and minimal feeder access. Used mainly to compare one placement system to another.

Placement Repeatability A placement system's ability to continuously place a component at a designated X-Y coordinate location within a specified tolerance.

Positional Repeatability A positioning system's ability to continuously position itself at a designated X-Y coordinate location within a specified tolerance.

Servo Motor A drive motor, operated by the output of a servo amplifier, that advances continuously.

Stepper Motor A drive motor that advances in uniform increments.

7.0 INTRODUCTION

Component placement is becoming the most mature part of the surface mount assembly process, probably because it has received the most attention over the years. Equipment availability ranges from inexpensive manual placement systems to very expensive high-speed and/or high-accuracy placement systems. Other terms have been used to describe placement equipment, including "pick and place" and "onsertion" (a carryover from autoinsertion equipment). The placement system is significant for two reasons: it is usually the most expensive equipment in a surface mount assembly line, with component feeders being a significant part of the investment, and it is typically the process that determines the output of a surface mount assembly operation (in terms of assemblies per hour).

Placement equipment has evolved and matured at a rapid pace. Early placement systems were slow, unreliable, and inaccurate. In less than ten years the equipment has become much faster, much more reliable, and much more accurate. Three important elements define a successful placement system: software, positioning mechanisms, and component feeders. Component centering is currently undergoing a major revolution. The original component centering method involved the use of mechanical fingers or chucks. Vision-based component centering is a new method that eliminates the mechanical mechanism. This method was originally developed for the placement of fine-pitch components, but it has improved the accuracy and flexibility of placement equipment used to place standard surface mount components as well.

One very confusing element of component placement is placement rate. All equipment suppliers provide a maximum placement rate for their equipment. This can be a very misleading figure. Some figures are very conservative, while others are very generous. They are usually based on optimum conditions, which are usually not achievable in an actual manufacturing environment. Many figures are based on very limited head and/or table movement and feeder access. The user needs to know what the *actual* placement rate is, which is the average placement rate that can be achieved on a continuing production basis. The actual placement rate is usually 50% to 85% of the maximum placement rate. Factors that influence placement rate are head and/or table travel distance (directly related to PCB size) and the number of different components per assembly (affects feeder access time).

Also confusing is the maximum number of feeders that can be loaded onto a placement system. Equipment suppliers usually express this figure using the 8mm tape and reel feeder. Other feeder types can use considerably more space, which significantly reduces the number of feeders that can be loaded onto the system.

This chapter will address standard, off-the-shelf placement systems and their important elements, including design and classification, positioning equipment, and component feeders. Custom and semicustom robotic work cells have also been used to place surface mount components; since they are primarily custom applications they are beyond the intended scope of this text.

7.1 PLACEMENT EQUIPMENT CLASSIFICATION

Placement equipment can be classified in several different ways: by design (gantry versus turret), by placement rate (low speed versus high speed), by cost (inexpensive versus expensive), or by functionality (flexible versus dedicated). Many of these classifications overlap one another. The clearest way to classify placement equipment is by design and functionality.

7.1.1 Placement Equipment Design

Placement systems are designed using one of two concepts: an overhead X-Y gantry or a rotary turret with an X-Y table. There are many variations, but most placement systems are designed using one of these two methods.

Gantry Systems
The basic X-Y gantry design is the most common system in use today, primarily because of the large market for low-cost, flexible placement systems. This concept is also preferred for fine-pitch component placement because it is very accurate and it does not move the PCB after each placement, which may shift a fine-pitch component out of position. A typical X-Y gantry system consists of a single overhead main gantry, which contains the placement head (see section 7.2), two parallel support gantries that are perpendicular to the main gantry, a tooling plate or edge-hold conveyor that positions and secures the PCB during placement, and feeder blocks that position and secure the component feeders (see section 7.5). Options may include a tool changer for vacuum nozzles and mechanical and/or vision centering mechanisms (see section 7.4). Some systems also have vision recognition and adhesive dispensing capability mounted on the placement head. A matrix-tray handling system for fine-pitch components can be added to more advanced equipment. Some of the more advanced equipment can also handle feeders that cut and form the leads of fine-pitch molded carrier-ring components prior to placement. An illustration of a typical single-head X-Y gantry system is shown in Figure 7.1.

The base frame supports the gantry and may contain the tooling plate, edge-hold conveyor, X-Y table and feeder blocks. It may also contain the controller and one or two video monitors. The tooling plate consists of a flat metal plate that positions the PCB with tooling pins or edge registration. Edge-hold conveyors transfer the PCB into and out of the system automatically using a belt or chain conveyor. Edge-hold conveyors also use tooling pins or edge registration to position the PCB. Some systems use a push plate to lift the PCB off the edge-hold conveyor and position it against a reference

118   Applied Surface Mount Assembly

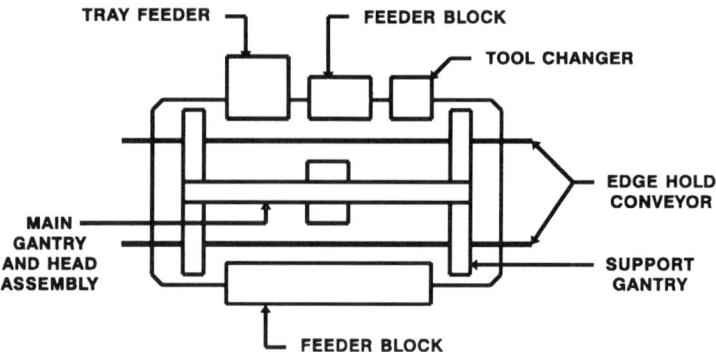

FIGURE 7.1.   X-Y Gantry Placement System Concept.

plane. At least one system uses a table that moves in the Y direction only, while the gantry only moves in the X direction. Tooling pin registration is more accurate and repeatable than edge registration because tooling holes can be drilled in a PCB more repeatably than a PCB edge can be routed. Use of a vision recognition system (see section 7.4) makes this issue less of a concern. Feeder blocks are located around the placement area. Most systems mount the feeder blocks on the front and back of the system, as shown in Figure 7.2. Some systems mount the feeders only on the back, while others mount them on all four sides. Gantry systems can handle multiple component feeder configurations, including tape and reel feeders, vibratory feeders, gravity feeders, and tray feeders. The maximum number of feeders a system can support ranges from approximately 20 to 100 8mm tape and reel feeders. If wider feeders are used the number decreases.

Because in most designs only one placement head is available, gantry systems are considerably slower than multiple-head rotary turret systems. However, there are some gantry systems with multiple placement heads that are reasonably fast. Maximum placement rates for single-head systems range from approximately 1,000 to 6,000 components per hour, while maximum placement rates for multiple-head systems vary considerably depending on the configuration. Typical systems place approximately 10,000 to 14,000 components per hour.

Gantry systems can be quite inexpensive. Manual tabletop models, as pictured in Figure 7.3, are available for approximately $6,000 without component feeders, while automatic tabletop models are priced around $50,000 without component feeders. The more advanced, fully equipped systems, such as the one shown in Figure 7.4, can cost $400,000 without component feeders.

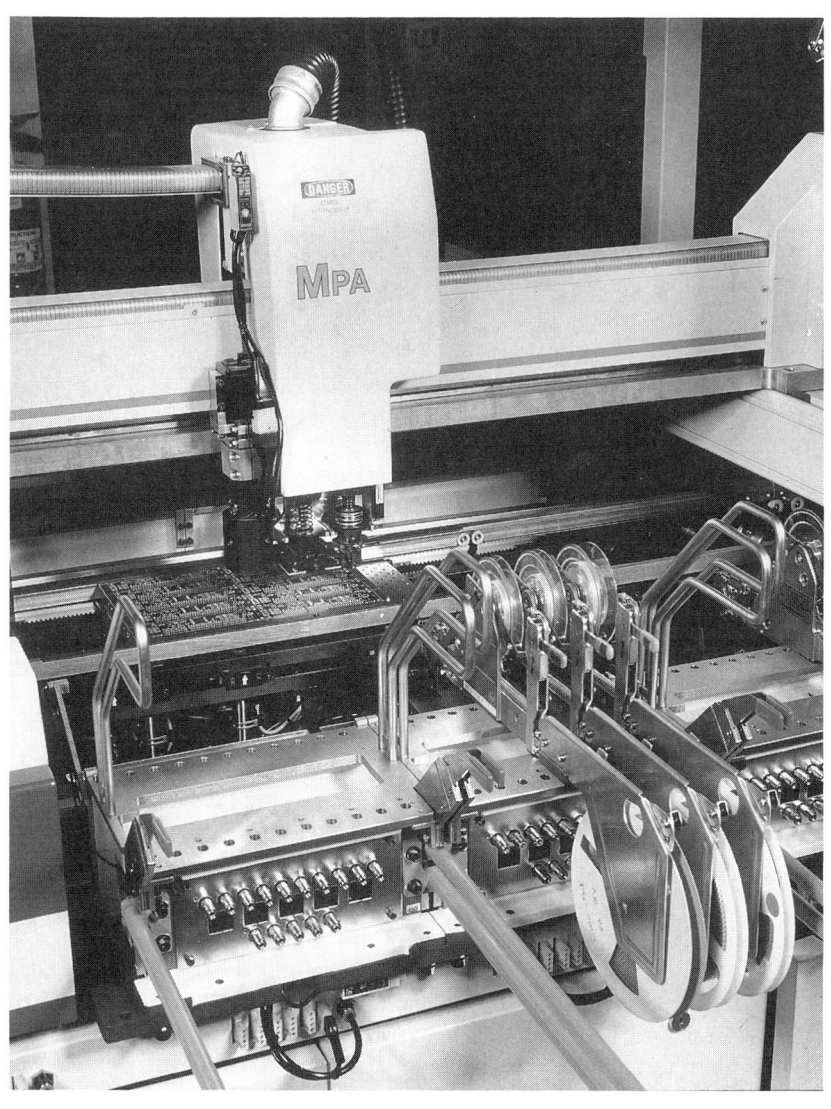

FIGURE 7.2.   Component Feeder Blocks on a X-Y Gantry System.

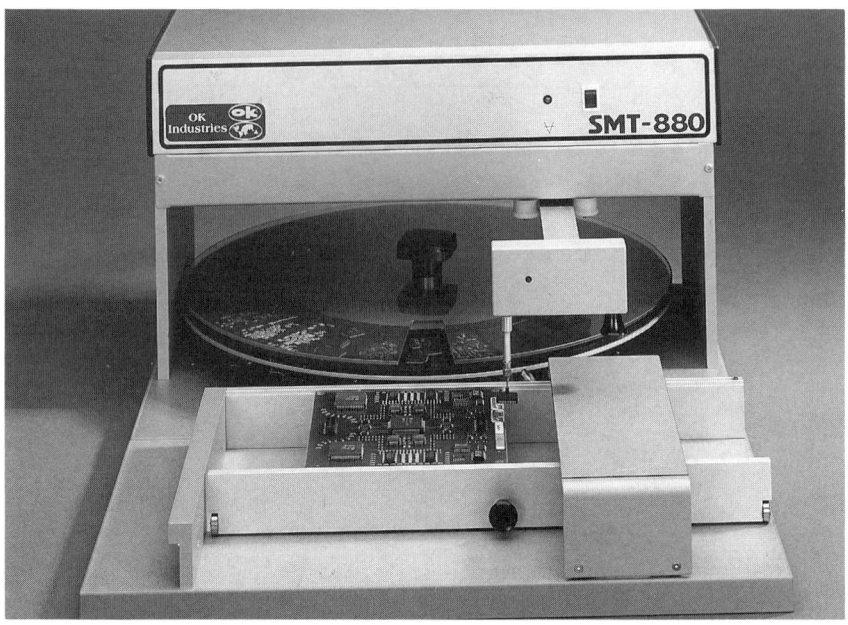

FIGURE 7.3.   Example of a Manual Placement System.

Rotary Turret Systems

The rotary turret design, as shown in Figure 7.5, has become very popular for high-speed placement systems. The intended market for these systems is dedicated, high-speed placement. However, they can be used successfully in lower volume applications as well. A rotary turret system consists of a main turret, an X-Y table, an edge-hold conveyor that positions and secures the PCB, and a linear carriage that positions and secures the component feeders. The main turret may contain individual heads (usually 10 or 12). Each head contains from 1 to 5 vacuum nozzles, as shown in Figure 7.6. Several designs use one main turret with approximately 60 to 100 vacuum nozzles. Component centering (see section 7.4) is accomplished using mechanical mechanisms or vision recognition. A rotary turret system is shown in Figure 7.7.

As with gantry systems, the base frame supports the main turret, as well as the edge-hold conveyor, X-Y table, and the feeder carriage for the component feeders. It may also contain the controller and one or two video monitors. Because all of these systems are intended for high-volume production, they all use edge-hold conveyors to automatically transfer the PCB into and out of the system using a belt or chain conveyor. The edge-hold conveyor

FIGURE 7.4.   Example of X-Y Gantry Placement System.

uses either tooling pins or edge registration to position the PCB. Some systems use a push plate to lift the PCB off the edge-hold conveyor and position it against a reference plane. As noted before, tooling pin registration is more accurate and repeatable than edge registration. Use of a vision recognition system (see section 7.4) makes this issue less of a concern. The feeder carriage moves back and forth behind the turret to position the component feeders, as shown in Figure 7.8. A unique feature on some systems is a split feeder carriage. This allows the carriage to be operated in several ways. The carriage can be joined together and operated as a single unit, or it

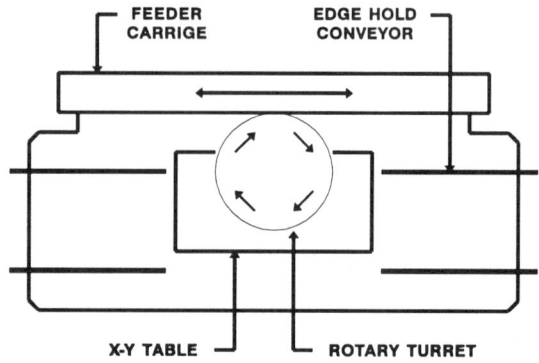

FIGURE 7.5. Rotary Placement System Concept.

FIGURE 7.6. A Rotary Turret Placing Components.

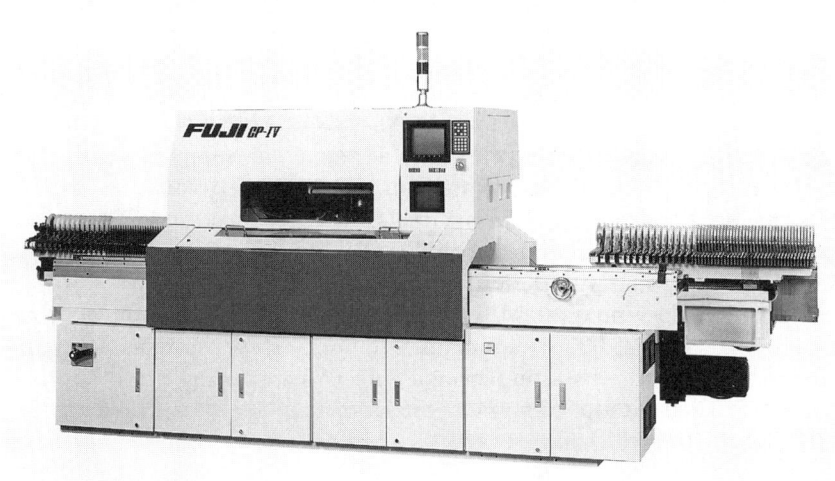

FIGURE 7.7. Example of a Rotary Turret Placement System.

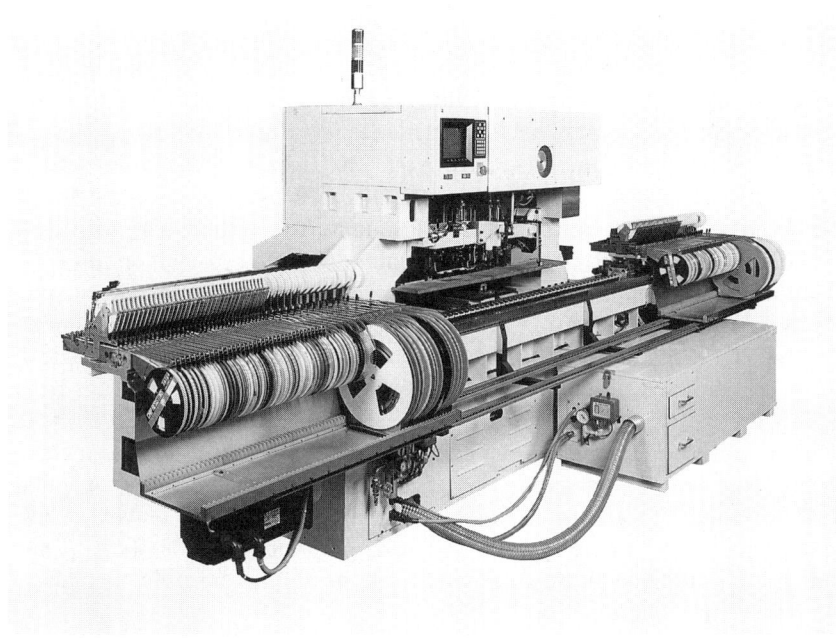

FIGURE 7.8. Compent Feeder Carriage on a Rotary Turret System.

can be split and operated as two individual units. The advantage to operating the carriage in two sections is set-up. While one section is placing components, the other section is moved off-line and the feeders for the next job are set up. This allows the system to be down for only a few minutes between jobs. The maximum number of feeders a system can support ranges from approximately 100 to 300 8mm tape and reel feeders. If wider feeders are used the number decreases. Tape and reel feeders (see section 7.5.3) are used exclusively because they are the only feeder type that has the speed and reliability to keep up with these systems.

Maximum placement rates for rotary turret systems range from approximately 14,000 to 25,000 components per hour. These systems are fairly expensive, with prices starting around $300,000 and going up to approximately $500,000 without component feeders.

7.2 PLACEMENT HEADS

Placement heads come in many different configurations depending on the type of equipment used (gantry versus turret). Each placement head consists of one or more vacuum nozzles, some form of tactile sensing, one or more spindles, and possibly a centering mechanism, depending on whether centering is accomplished internally or externally (see section 7.4).

The placement head is responsible for removing the component from the feeder, orienting it correctly, and placing it on the PCB. The component is picked up and transported by a vacuum nozzle. Nozzles cover a limited range of component types and sizes. Usually three to five different nozzles sizes can cover a full range of components from 0805 to PLCC84. Vacuum or optical sensing is used to verify that the component is actually in place on the nozzle tip. If the component is lost before placement, the system will institute a repair function that will obtain another component and place it on the PCB. This is a very important feature. Tactile sensing is also very important because it prevents a component from being crushed between the placement nozzle and the PCB. In simple terms, tactile sensing controls the Z-axis (vertical) stroke of the nozzle. Tactile sensing tells the nozzle when contact has been made with the PCB. The nozzle then releases the component and retracts to its home position. Tactile sensing can be provided by spring-loaded mechanisms, load cells, programming, or a combination of all three. Angular position (theta) is accomplished by rotating the nozzle. Less expensive systems may only be able to rotate the nozzle to four positions (0°, 90°, 180°, and 270°), while the more expensive systems should be able to place a component in 1° increments.

As noted before, most gantry systems have a single placement head. The more advanced systems incorporate tool changers that can select different nozzles and centering mechanisms (if internal centering is used) so a full range of components can be handled. Rotary turret systems have multiple nozzles available on the turret. Tool changers are not employed on these systems. The turret is rotated until the proper nozzle is in position. All component centering is accomplished externally.

7.3 POSITIONING SYSTEMS

The positioning system—either the X-Y gantry, the X-Y table, or the rotary turret—is directly affected by the drive mechanism, the drive motor, and the verification method.

Typical drive mechanisms are cables, timing belts, lead screws, and ball screws. Cables and timing belts are the least accurate and repeatable, while lead screws and ball screws are the most accurate and repeatable. Factors such as wear, material expansion rate, and backlash can and do affect their performance.

Drive motors consist of two types: stepper motors and servo motors. A stepper motor advances the drive mechanism in uniform increments until the correct position is achieved. A servo motor continually advances the drive mechanism until the correct position is achieved. Stepper motors are slower and less accurate than servo motors. Servo motors are faster and more accurate than stepper motors, and they can handle larger loads. Stepper motors are less expensive than servo motors.

Encoders are typically used to verify position. Basically, two types of encoders are used: rotary and linear. Stepper motors use either rotary or linear encoders that count pulses, while servo motors use digital rotary encoders that read a binary number. In general, rotary encoders tend to have good repeatability and an excellent mean time between failures (MTBF), while linear encoders tend to have excellent repeatability and a good MTBF. Linear encoders have an advantage because they are closer to the end result. Rotary encoders can be influenced by backlash in the drive mechanism; however, good positioning systems are designed to compensate for the backlash.

7.4 COMPONENT CENTERING

When a placement program is written, the X-Y coordinates are based on the distance from program zero (usually the center of one of the tooling holes) to the center of the vacuum nozzle. If a component is to be placed correctly the centroid of the component must be in alignment with the center of the nozzle, or the distance between the center of the nozzle and the centroid of the component (offset) must be known. Since the component feeder does not provide the component with a precise position prior to pickup by the nozzle, the placement system must furnish some method of nozzle to component alignment. Alignment can occur internally or externally, as indicated in Figure 7.9. Internal centering means the centering mechanism is located on the placement head, while external centering means the centering mechanism is located on the frame of the placement system. The original component centering systems were mechanical mechanisms, and they are still widely used. More advanced placement systems use vision-based centering, especially for fine-pitch component placement.

7.4.1 Mechanical Centering

A mechanical-based component centering system attempts to reposition the component on the nozzle after it has been removed from the component feeder. This is accomplished by pushing on the component from all four sides with a mechanical centering mechanism. This mechanism consists of four spring-loaded jaws or fingers that close around the component and move it in to the proper position. After repositioning, the component centroid should be in alignment with the center of the nozzle. Mechanical centering is a reliable centering method for most components.

Centering can occur internally or externally. Many different concepts have been used for mechanical component centering. Some have been more successful than others. This is an important element to consider when selecting a placement system. Poorly designed and/or maintained mechanical centering mechanisms can easily damage components. Mechanical centering is also subject to the tolerance and variations in the components themselves. This is not a concern with components that have a large placement tolerance (for example, a 1206 resistor on a 1206 land pattern).

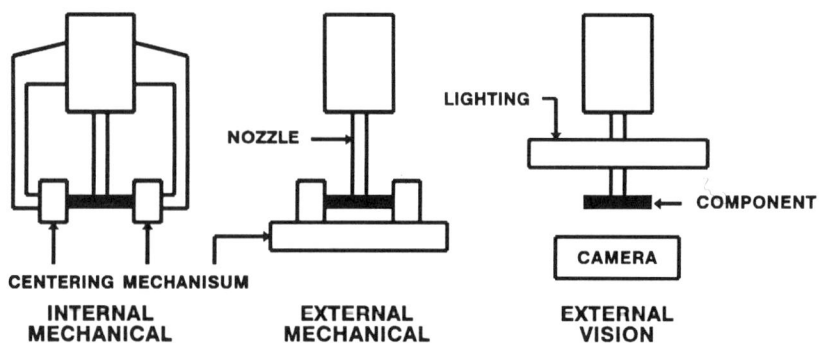

FIGURE 7.9. Component Centering Methods.

However, complex components, such as a large PLCC, are more difficult to center mechanically. Fine-pitch components are very difficult, if not impossible, to center mechanically. In addition to centering problems, the fine-pitch leads may be damaged by the mechanical centering mechanism.

7.4.2 Vision Centering

Vision component centering is an external centering method. It is becoming a very useful and powerful tool in the placement process. Vision centering is a necessity for fine-pitch component placement. Two vision cameras are usually employed in the placement system: a downward looking CCD camera for fiducial recognition (see section 7.4.3) and an upward looking CCD camera for component recognition. Important camera-related issues are its field of view (FOV) and its resolution. Vision systems that can view all or most of the component and/or leads usually operate faster than systems that must compile a complete view from several subviews. However, the resolution of limited view systems is normally better. The resolution is usually directly related to the number of pixels in the array. More pixels provide better resolution. Some systems use subpixel processing, which improves resolution further, but at the cost of increased processing time.

Selecting the right vision system for the application is very important. Two types of vision systems are used: binary and gray scale. Both methods use vision alignment in the same fundamental manner. The vision system determines the X, Y, and theta offset of each component prior to placement. In addition to determining the component offset, the vision system can also inspect the component for dimensional integrity and lead damage (skew and coplanarity problems).

Binary imaging, which is the older of the two methods, locates a feature using the contrast between black and white images. The vision system converts a gray-scale image to a binary (black and white) image, which is referred to as thresholding. The system then uses an algorithm to determine the centroid of a white feature that is placed against a black background. Binary imaging has been popular because it requires only a moderate amount of computing capability; however, it is sensitive to contrast and lighting changes. Binary imaging is successful on components down to a lead pitch of 0.5mm (0.0197″). For finer pitch components it will probably be necessary to use gray-scale imaging.

Gray-scale imaging operates in a manner similar to binary imaging. The important difference is that a gray-scale image is used, rather than a binary (black and white) image, thus more detail is seen by the vision system. A gray-scale image is normalized, which means it is not as sensitive to contrast and lighting changes as a binary image. The drawback to gray-scale imaging has been its excessive processing time. But, recent improvements in computers and software have made this problem less significant.

7.4.3 Fiducial Recognition

Vision alignment on placement systems is becoming an important method for neutralizing PCB fabrication errors. The vision system, either binary or gray scale, uses pattern recognition to locate fiducial marks or other features on the PCB. The vision system is provided with theoretical fiducial information. After a PCB enters the placement system the vision camera looks at the fiducial and compares the result to the theoretical value. The required offset is determined, and the placement system is adjusted accordingly.

Two types of fiducials are used in the placement process: global (PCB level) and local (component level). Usually two PCB-level fiducials are placed at diagonally opposite corners. This helps eliminate X, Y, and theta errors. Component fiducials are used with fine-pitch components to eliminate local variations. Some systems use one fiducial at the center of the land pattern, while others prefer two fiducials outside of the land pattern. Component fiducials and vision centering are used to precisely match the fine-pitch component to the land pattern.

Many patterns can be used for a fiducial, but a circle is a very good choice because it is symmetrical about its axis and thus is not susceptible to rotational offset error. Equipment suppliers usually provide recommendations for fiducial patterns. See Chapter 4 for more information on fiducial design.

7.5 COMPONENT FEEDERS

Component feeders, as shown in Figure 7.10, are one of the most critical elements in a placement system. An otherwise excellent placement system will be a failure if poorly designed and/or manufactured feeders are used. Successful feeders incorporate the following:

- They use a minimum amount of space on the placement system.
- They provide smooth, continuous component feeding.
- They allow easy loading of components onto the feeder.
- They provide precise positioning of components for pickup.
- They prevent mechanical and electrical component damage.
- They provide maximum reliability and minimum maintenance.

And they must accomplish all of the above at a reasonable price. Component feeders can be divided into five categories: gravity, vibratory, tape and reel, tray, and cut and form.

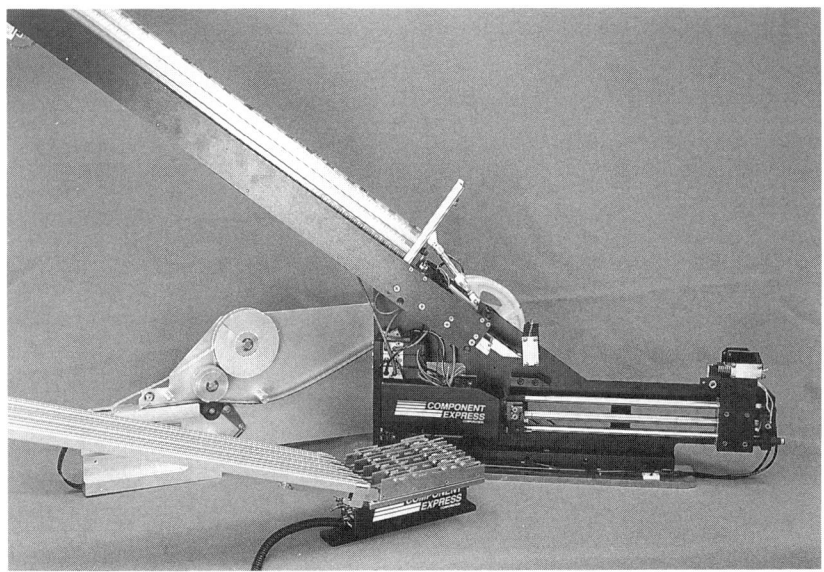

FIGURE 7.10. Component Feeders: Tape and Reel, Vibratory and Gravity (Left to Right).

7.5.1 Gravity Feeders

Gravity feeders are simple and relatively inexpensive since they rely upon gravity to feed the components from the tube to the pickup point. Each feeder consists of a base and one or more feeder tracks. They require less space than tape and reel feeders. This allows a gravity feeder to provide more inputs than a tape and reel feeder in the same amount of space. The plastic tube containing the components is mounted to the back of the feeder at a steep incline. The components slide down the tube and into the feeder track. One component at a time is positioned at the pickup point. Multiple tube feeders can automatically insert and remove tubes.

Gravity feeders are primarily intended for low-volume applications. They usually require frequent operator intervention, either to add components or correct a misfed component. Even at a steep incline components will stick in the tube. Gravity feeders allow components to contact each other, which is a concern. Dual-sided components, such as the SOIC, may have problems with mold flash on the ends of the components. The mold flash from two different components can overlap each other in the feeder, which causes pickup problems. Lead damage can occur on four-sided components, such as the PLCC, because the leads on two sides are in contact with each other under pressure. This is especially a concern with larger components, such as the PLCC68.

7.5.2 Vibratory Feeders

Vibratory feeders are similar to gravity feeders, except the components are moved through the feeder by vibration rather than by gravity. As with gravity feeders, vibratory feeders require less space than tape and reel feeders. These feeders consist of a vibrating base and one or more feeder tracks. Some feeders insert a metal tube, containing the components, into the track. Plastic tubes may be mounted to the back of the feeder at an incline to feed the components into the vibrating track. Once they are on the feeder track, the components vibrate down to the pickup point.

Vibratory feeders are also primarily intended for low-volume applications. They have the same concerns that are associated with gravity-type feeders, namely frequent operator intervention to add components or correct misfeeds, and components that are in continuous contact with each other. In addition, if the component is vibrating excessively at the pickup point the nozzle may have difficulty picking up the component.

7.5.3 Tape and Reel Feeders

Tape and reel feeders are the most common and reliable feeders in use. These feeders index a carrier tape, containing the component, through the feeder to the pickup point. Carrier tapes, which provide the best component protection, will be discussed later in section 7.6. Most tape and reel feeders are a one-piece unit that consists of a real holder, a carrier tape indexing mechanism, and a cover tape peeling mechanism. Some systems have the ability to cut the used carrier tape into small pieces and place them into a waste container, while other systems simply index the carrier tape out the back of the feeder where the operator cuts off the used tape. The indexing mechanism can be activated either mechanically, pneumatically, or electrically. To prevent components from jumping out of the carrier tape during indexing the feeder should index the carrier tape first and then peel the cover tape off. Feeders are currently available to handle 8mm, 12mm, 16mm, 24mm, 32mm, 44mm, and 56mm carrier tapes.

Tape and reel feeders are suitable for any production volume, but they are particularly well suited for high-volume applications. One problem has been the availability of some components, especially integrated circuits, in tape and reel. Most component suppliers require a specific minimum order before they will supply the component in tape and reel. In some cases the minimum order will exceed the user's requirements.

7.5.4 Tray Feeders

Tray feeders have evolved primarily to support fine-pitch technology. The delicate nature of fine-pitch leads requires that the components be packaged in such a manner that the leads do not contact other components or the component packaging. Trays have also been used to feed other component types, primarily in low-volume applications. Tray feeders can be small and simple or large and complex. Their design varies between equipment manufacturers. A simple tray feeder may consist of a mounting platform in the feeder block that the placement head can access directly, while more complex systems may consist of a multiple-level tray feeder with a shuttle mechanism to deliver the component to the placement head.

7.5.5 Cut-and-Form Feeders

On-line lead cutting and forming feeders are expensive and complex. They are coming into use for very fine pitch components, such as molded carrier-ring devices. These feeders consist of an input mechanism that will feed the component from a plastic tube into the feeder and a cut-and-form mechanism that will cut the component from the carrier ring, form the leads into a gull-wing configuration, and discard the used carrier ring. They can only handle one component size at a time. These feeders require considerable space on the placement system.

7.6 COMPONENT TAPING

7.6.1 Taping Materials

Specifications for standard carrier and cover tapes—such as tape width, cavity size, component orientation, and peel strength—are documented in EIA-481 "Taping of Surface Mount Components for Automatic Placement." Currently, standard carrier tape widths are 8mm, 12mm, 16mm, 24mm, 32mm, 44mm, and 56mm. Carrier tapes consist of a cavity for each component and sprocket (drive) holes down one or both sides, depending on the width of the tape.

The 8mm and 12mm carrier tapes can be either punched or embossed, while the 16mm through 56mm carrier tapes are all embossed. Punched carrier tapes, which are usually the same thickness as the component, are made from paper or plastic, although paper is the most common. Cover tape is applied to both the top and bottom of a punched carrier tape. Embossed carrier tapes are made from plastic or metal, with plastic being the most common. Cover tape is only applied to the top of an embossed carrier tape. Plastic carrier tape is usually preferred over paper carrier tape because dust particles from the punched paper tape can impair mechanical mechanisms and vacuum nozzles. However, punched carrier tapes handle very small resistors and capacitors more effectively because of their superior cavity definition (straight, sharp walls).

7.6.2 Component Taping Systems

Some surface mount manufacturing operations have a limited in-house capability to tape and reel components. There are usually two reasons for this: the user does not have a high enough volume to purchase components in tape and reel or to cover emergencies when the components are delivered to the user in tubes, trays, or bags. Component taping systems are available with low-, medium- and high-volume capability. Low-volume and medium-volume systems are the logical choice for a limited in-house capability. High-volume dedicated systems are used strictly by component suppliers.

A low-volume manual system, shown in Figure 7.11, can typically tape and reel 300 to 2,000 components per hour. Manual systems are quite flexible because no dedicated tooling is used to load the component into the carrier tape. The operator simply places the components on a tray and loads them into the carrier tape using a vacuum pencil. These systems usually cost between $10,000 and $25,000.

FIGURE 7.11. Example of a Manual Component Taping System.

FIGURE 7.12. Example of a Automatic Component Taping System.

A medium-volume system, shown in Figure 7.12, can typically tape and reel 5,000 to 15,000 resistors and capacitors per hour and 1,000 to 8,000 integrated circuits per hour. Two different systems may be required to handle small and large components. These systems frequently use semiautomatic or automatic mechanisms to load the components into the carrier tape. It is important to maintain system flexibility. Tooling that is difficult to change and/or expensive to fabricate limits the effectiveness of the system. These systems usually cost between $35,000 and $75,000.

REFERENCES
EIA-RS-481. "Taping of Surface Mount Components for Automatic Placement," EIA, Washington, D.C. April 1991.
1. Brinton, James B. "Viewing Fine-Pitch Placement." *Circuits Manufacturing*, January 1990, pp. 20–26.
2. Clerici, Tom and Merrit, Merrill. "Tape and Reel Assembly." *Surface Mount Technology*, June 1991, pp. 20–26.
3. Martin, Chris. "Selecting Placement Equipment for Surface Mount." *Surface Mount Technology*, June 1990, pp. 19–23.
4. McWater, Kathy. "Future Trends for Vision Systems in Surface Mount Placement." *Surface Mount Technology*, April 1989.
5. Peterson, James R. "Component Feeders for SMT Board Assembly." *Surface Mount Technology*, June 1990, pp. 31–35.

6. Prasad, Ray P. Surface Mount Technology—Principles and Practice. New York: Van Nostrand Reinhold, 1989.
7. Silver, William M. "Gray-Level Processing in Machine Vision." *ESD*, May 1987.
8. Smith, Fred. "SMT Equipment: Simplifying the Critical Choice." *Circuits Assembly*, September 1991, pp. 40–45.

8
Reflow Soldering and Adhesive Curing

GLOSSARY

Conduction   The transfer of heat through a material.
Convection Reflow   A reflow system that transfers greater than 50% of its energy (heat) to the printed circuit board by convection.
Convection   The transfer of heat by the circulation of a fluid or gas.
Infrared (IR) Reflow   A reflow system that transfers greater than 50% of its energy (heat) to the printed circuit board by radiation.
Opacity   The ability of a surface to impede the transmission of radiation.
Profile   The relationship of time versus temperature during the reflow process. Each reflow method will have a preferred time/temperature profile.
Radiation   The emission of energy (heat) from an object.
Reflectivity   The extent to which a surface reflects radiation.
Reflow Temperature   The temperature at which proper wetting and solder joint formation takes place. Usually stated at 25°C (77°F) above the melting temperature of the solder alloy.
Thermocouple   A temperature measuring device, which is manufactured using two different metals that are joined together at one end to form a junction. The junction, when heated, generates a small thermoelectric voltage. The voltage change represents a change in temperature.
Tomstoning   A rectangular or cylindrical component that has flipped into a vertical position during the reflow soldering process. Caused by unequal solder surface tension. To prevent tomstoning the surface tension forces must be equal on both sides of the component.
Translucency   The partial transmission of radiation through a surface or material.

Transparency The unimpaired transmission of radiation through a surface or material.
Vapor-Phase Reflow A reflow system that transfers greater than 50% of its energy (heat) to the printed circuit board by condensation phase change.
Zone An area of uniform temperature or heating rate that is controlled by a continuous loop of temperature sensing and power adjustment. An eight-zone reflow system would have four zones on the top and four zones on the bottom.

8.0 INTRODUCTION

Reflow soldering may appear to be a simple process. It can be accomplished in something as simple as a small bench-top convection oven. However, looks can be deceiving! Reflow soldering is actually a very complex process with many variables. All mass reflow systems incorporate convective, conductive, and radiant means of heat transfer. Various methods are used to achieve reflow, but they all strive to achieve the same fundamental results at certain points in their process. Five different phases take place during the reflow process. A reflow system must incorporate the following:

1. Evaporate solvents from the solder paste
2. Activate the flux and allow fluxing action to occur
3. Carefully preheat the components and printed circuit board (PCB)
4. Melt the solder and allow wetting of all joints
5. Cool the completed assembly at a controlled rate to an acceptable temperature

The reflow system must also dispose of fumes generated during the heating process. Overall, it is a very complicated process to establish and control!

Adhesive curing, which is a much simpler process than reflow soldering, will also be discussed in this chapter. Most reflow equipment can be used successfully for adhesive curing.

Reflow equipment has changed more frequently than any other surface mount assembly equipment. During the last ten years four different design concepts have emerged: vapor phase, lamp infrared (IR), panel infrared (IR), and most recently forced convection. Vapor-phase reflow technology evolved first and was the method of choice for several years. But the early vapor-phase systems were plagued by maintenance problems and concerns about operating costs. Although not trouble free either, infrared systems, once they matured, became the preferred approach. Today, panel infrared systems, in one form or another, are the most common equipment type in use. The latest development in reflow technology, forced convection, is rapidly gaining acceptance and will certainly influence future equipment.

Throughout the industry there are advocates and supporters of each method.

The focus of this chapter is to review the common mass reflow methods used in the surface mount industry today. More specialized reflow processes, such as the laser, will not be covered. Because there are so many different designs available today, reflow equipment is becoming more difficult to classify. The reflow methods discussed in this chapter will be separated into four categories: vapor phase, lamp IR, panel IR, and forced convection. This classification is based on the method used to achieve reflow. Because so many different equipment types and designs are available there may be some overlap between classifications. This chapter will also review other important reflow issues including time/temperature profiles, control systems, transport concepts, controlled atmospheres, profiling methods, and adhesive curing.

8.1 PROFILING ISSUES

8.1.1 Time/Temperature Profiles

A time/temperature reflow profile should be developed before determining the reflow method or selecting equipment. Many factors influence a reflow profile. Each one should be clearly defined and understood. Examples of generic profiles are provided by Figures 8.1, 8.2, and 8.3.

The reflow process, or profile, can be described in three phases: preheat, reflow, and cool down. The preheat phase should be viewed as the preparatory phase. All actions leading up to proper reflow occur during preheat. During this phase the PCA is heated at a controlled and uniform rate, solvent evaporation commences, metal oxides are removed by the activated flux, and the solder particles begin to melt. The reflow phase occurs when the temperatures of the solder and the solderable surfaces are above the melting temperature of the solder alloy. This elevated temperature is required to reduce surface tension and promote proper wetting of the surfaces to be soldered. The amount of time the solder is above its melting temperature, or dwell time, is a significant factor. The cool-down phase helps control the dwell time and provides the proper cool-down rate for the solder joint, which promotes proper grain structure formation.

8.1.2 Printed Circuit Boards

Printed circuit boards can be damaged by excessive exposure to heat. The glass transition temperature (Tg) of the PCB material must be known to help define the reflow profile. Common epoxy/glass materials have a Tg between 120°C (248°F) and 130°C (266°F) (material suppliers can provide accurate

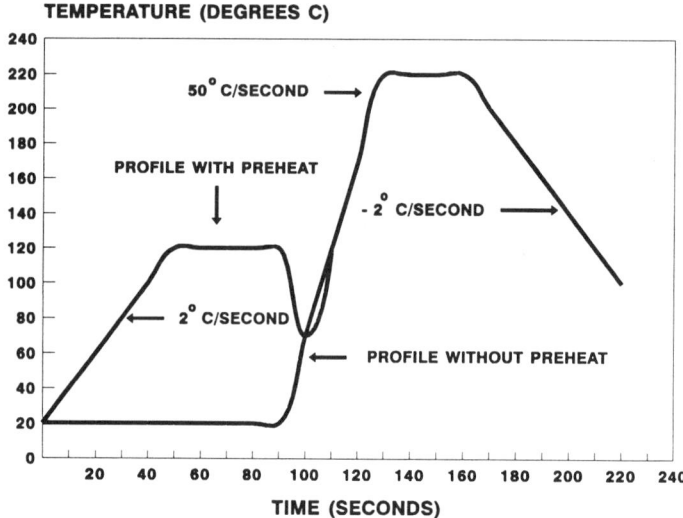

FIGURE 8.1. A Common Vapor-Phase Time/Temperature Profile.

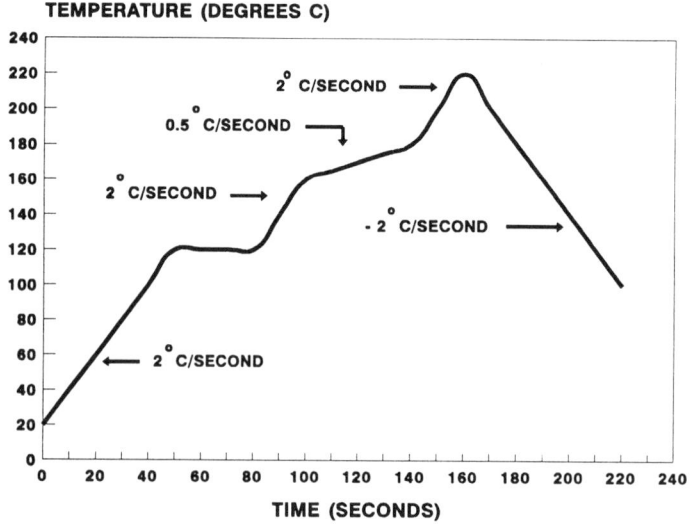

FIGURE 8.2. A Common Infrared Time/Temperature Profile.

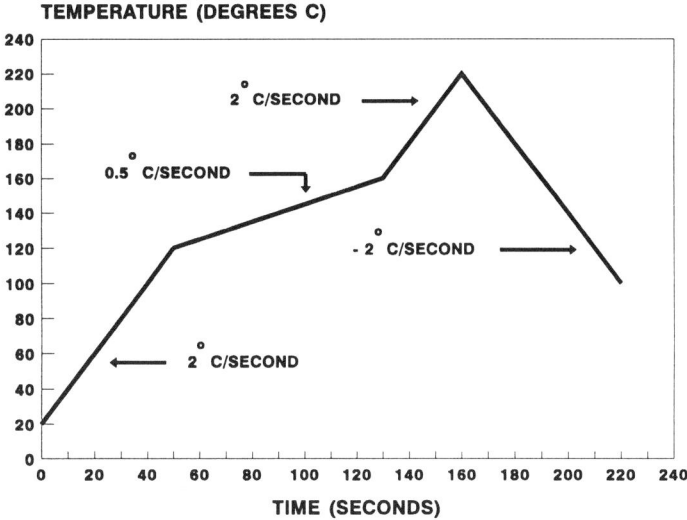

FIGURE 8.3. A Common Forced Convection Time-Temperature Profile.

figures). Two potentially damaging things occur above Tg. The material becomes plastic, so the PCB becomes soft and loses its rigidity. If an edge-hold conveyor is used this loss of rigidity may cause the PCB to fall off the conveyor, especially if break-off tabs have been used. Also, the expansion rate of the material increases dramatically. This increased expansion rate can be very damaging, especially in the Z axis. This damage can affect the integrity of the PCB, particularly the plated through-holes. It is best to limit the time above Tg to 120 seconds.

The surface finish on the PCB lands will have a major impact on solderability. For information on PCB land finish refer to Chapter 3.

8.1.3 Flux

Flux has two attributes that affect the reflow process. To properly remove oxides it is important to understand the activation temperature and activation time of the flux. A common mistake is to use a time/temperature profile that consumes the activator before the solder melts. On the other hand, it is important to have the flux active long enough to remove the oxides from the PCB, component leads, and the solder paste powder. Ideally, the last of the activator would be consumed just as the solder begins to melt. An acceptable activation time for most flux material is a minimum of 30 seconds and a maximum of 90 seconds. Flux usually becomes active at 110°C to 120°C (230°F to 248°F).

8.1.4 Components

Components can be damaged by the incorrect application of heat. All components have a limit as to the amount of heat they can be exposed to without damage. Most components can tolerate the reflow temperature range of 210°C to 220°C (410°F to 428°F) for 20 to 60 seconds without damage. The component supplier can provide details about maximum time and temperature limitations.

Thermal shock, caused by a rapid temperature increase, can crack and delaminate components, especially capacitors. In most cases, capacitors are the limiting factor on the PCA. The general rule has been not to exceed a 2°C/second (3.6°F/second) temperature increase or decrease. Some recent information has stated that temperature increases from 3 to 6°C/second (5.4 to 10.8°F/second) are safe to use. It is important to understand what the component supplier recommends and why. Capacitor manufacturers still advise limiting the temperature increase to 2°C/second (3.6°F/second).

Recent data indicates that moisture trapped inside integrated circuit packages may contribute to package cracking. The cracking occurs as a result of moisture expansion during the reflow (heating) process. Protective dry packaging and baking can be used to prevent this damage. Consult with the component supplier to determine if special packaging or baking is required. Refer to IPC-SM-786, "Recommended Procedures for Handling of Moisture-Sensitive Plastic IC Packages," for more information.

Component lead finish, which will affect solderability, is critical, particularly when determining which flux type to use. Information on component lead finish can be found in section 5.6.

8.1.5 Solder

Reflow temperature is usually 25°C to 40°C (77°F to 104°F) above the melting temperature. It is important to achieve this temperature, which allows the solder to wet the base metal surfaces properly. The time above the melting temperature, typically 20 to 60 seconds, is also significant because it allows the solder enough time to properly wet the base metal surfaces.

Cooling affects the final strength and integrity of the solder joint. In general, solder joints that are cooled at a reasonable rate achieve a small, fine-grain structure. This grain structure provides a stronger, more reliable solder joint. Cool-down rates of 1 to 2°C/second (1.8 to 3.6°F/second) are preferred; however, cooling rates up to 5°C/second (9°F/second) have been used.

8.2 VAPOR-PHASE REFLOW

Vapor-phase reflow soldering was developed by Western Electric Company in the early 1970s to solder wire-wrap pins in back panels. Soon after, the concept was adapted to reflow surface mount assemblies. The basic operating principle is fairly simple. A vapor-phase system generates a vapor zone by heating a stable, inert fluid to its boiling point, which is usually 25°C to 40°C (77°F to 104°F) above the melting temperature of the solder alloy. The vapor temperature is equal to the boiling point of the fluid. Operating temperature is determined by the boiling point of the fluid. To achieve reflow, a PCA is placed into the vapor zone. The vapor immediately condenses on the PCA and quickly elevates it to the proper reflow temperature. Heat transfer is independent of mass, size, and shape; however, the system must have enough heating capacity to continuously regenerate the vapor zone. All vapor-phase systems control the reflow process by controlling dwell time. The rate at which heat is transferred to the PCA is directly proportional to the difference in temperature between the vapor and the PCA.

Vapor phase was originally the most common method for reflowing surface mount assemblies. However, in recent years it has become less popular, as infrared technology and convection reflow technology have improved. Vapor-phase equipment has evolved into two types of systems: batch and continuous.

8.2.1 Vapor-phase Fluid

All currently available vapor-phase fluids are based on fluorinated compounds. These very stable, nonpolar fluids are colorless, odorless, and nonflammable. Contrary to popular belief, they do not provide any cleaning capability. They can, however, make cleaning easier because they prevent flux charring. They are not associated with chlorofluorocarbons (CFC), and they do not, based on current knowledge, damage the environment.

The fluids are available from a number of suppliers, with a temperature range from approximately 160°C to 260°C (320°F to 500°F). The following fluid factors are important:

Boiling Point—The temperature at which the fluid vaporizes.
Heat of Vaporization—The amount of heat required to convert one gram of
  the fluid to a vapor without increasing its temperature.

Molecular Weight—The sum of the atomic weights of the individual atoms of the fluid.

Pour-point Temperature—The lowest temperature at which the fluid will pour easily.

Surface Tension—A characteristic of a fluid that allows its surface to contract.

Thermal Conductivity—The heat conducting capacity of the fluid.

Vapor Pressure—The pressure exerted by the vapor when it is in equilibrium with its liquid form.

Viscosity—The resistance of the fluid to flow.

Under certain conditions the fluid can generate harmful vapors and corrosive acids. A very toxic substance, perfluoroisobutylene (PFIB), is created in the vapor by the thermal decomposition of the fluid. And when PFIB is combined with moisture it will form hydrofluoric acid (HF). Thermal decomposition is usually caused when the flux from the solder paste is deposited on the surface of the fluid heating elements. The deposit chars and hardens, creating an insulator on the heating element, which causes the element to overheat, and thus overheats the fluid. A method of on-line fluid filtering is required to remove flux from the vapor and fluid chamber. A good reliable ventilation system is also necessary to remove any harmful vapors that may form. A sound proactive maintenance program is vital.

8.2.2 Batch Systems

Batch systems, as shown in Figure 8.4, are intended for use in low-volume or laboratory environments. Their processing rate is much lower than continuous systems. They also differ from continuous systems because the PCA enters and exits from the same point. Figure 8.5 shows a batch vapor-phase system. Typical batch systems have a fluid/vapor chamber, a preheat zone (optional),

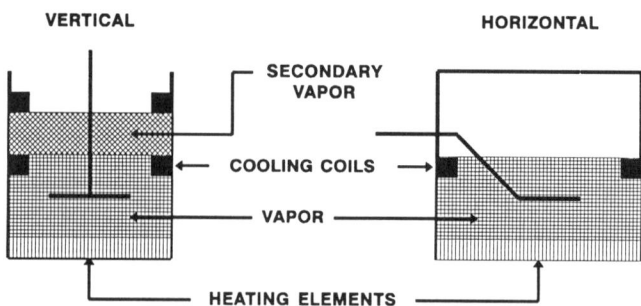

FIGURE 8.4. Batch Vapor-Phase System Concept.

Reflow Soldering and Adhesive Curing 143

FIGURE 8.5. Example of a Batch Vapor-Phase System.

a cool-down zone (optional), and a transport system. The fluid/vapor chamber contains heating elements, cooling coils, and of course the fluid and vapor. Some designs also use a secondary vapor zone, containing a chlorofluorocarbon (CFC), above the primary vapor zone to help contain the primary vapor. New materials are reportedly being developed to replace the CFC material. The profile is controlled by varying dwell time in the vapor zone.

The original batch systems used a vertical, overhead crane supporting a wire basket to transport the PCA into and out of the primary vapor zone. This concept uses a secondary vapor zone above the primary vapor zone. A newer design uses a horizontal edge-hold conveyor to transport a pallet containing the PCA into and out of the vapor zone. The pallet moves down an incline, but remains horizontal, into the vapor zone, and then returns along the same path. This design does not use a secondary vapor zone.

8.2.3 Continuous Systems

Continuous, or in-line, systems (as illustrated in Figure 8.6) are intended for use in medium- and high-volume applications. The PCA enters one end and

144   Applied Surface Mount Assembly

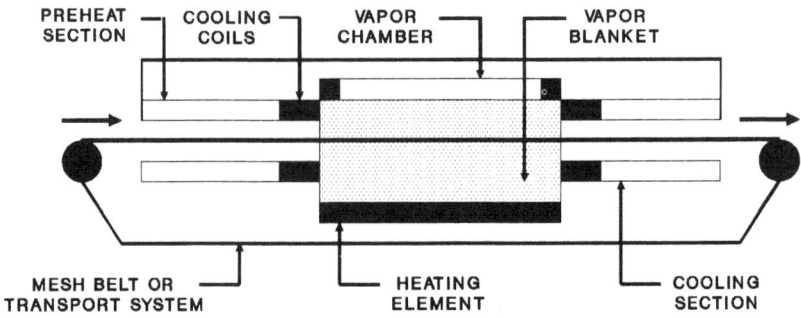

FIGURE 8.6.   Continuous Vapor-Phase System Concept.

exits from the opposite end. Standard in-line systems (as shown in Figure 8.7) consist of a fluid/vapor chamber, a preheat zone (optional), a cool-down zone (optional), and a transport system. The fluid/vapor chamber contains heating elements and cooling coils, along with the fluid and vapor. The profile, controlled by dwell time in the vapor zone, is adjusted by changing the transport speed.

FIGURE 8.7.   Example of a Continuous Vapor-Phase System.

The original in-line design uses a slightly inclined stainless steel mesh belt, usually 18 inches wide, to transport the PCA through the system. The mesh belt is pulled through the system by drive sprockets located on the exit end. Correct belt tension is crucial for smooth operation.

A different design uses an edge-hold conveyor that transports a pallet containing the PCA. The pallet moves down an incline, while remaining horizontal, into the vapor zone, and then moves up another incline, again remaining horizontal, out of the vapor zone. Multiple pallets can be loaded into the system.

A closed loop system that can monitor and control the mesh belt or edge-hold conveyor speed is a desirable option.

8.2.4 Preheat

It has been established that preheat is very beneficial to the vapor-phase reflow process. Preheating minimizes thermal gradients across the PCA, which helps to reduce wicking and tomstoning problems. Preheat can be provided by an internal or external source. The internal source is preferred because less heat is lost before the PCA enters the reflow zone. When a PCA is preheated externally most of the heat is lost before it reaches the reflow zone. The most common heater is an infrared (IR) panel (see section 8.3).

Vapor-phase Attributes

Vapor-phase systems provide equilibrium heating. The temperature difference between the PCA and the vapor is near zero.

Vapor-phase systems provide almost uniform heating. Heat is transferred quickly and efficiently from the vapor to the PCA.

Vapor-phase systems have a low source temperature. The vapor temperature remains constant. The reflow process and the PCA can not exceed the temperature of the vapor.

Vapor-phase systems provide a controlled environment. The vapor displaces other gases and provides an inert reflow environment.

Vapor-phase systems are easy to profile. The primary concern in vapor-phase profiling is dwell time. This is controlled by changing the speed of the transport system.

Vapor-phase systems can change profiles quickly. A change in transport speed, which increases or decreases dwell time, is all that is required to alter a profile. The system does not have to heat up or cool down.

Vapor-phase Concerns

Vapor-phase systems transfer heat rapidly. There is concern about the potential for component damage caused by thermal shock due to the rapid heating of this process, which can range from 15 to 50°C/second (59 to 122°F/second). Preheat is recommended to decrease thermal shock and to properly remove volitiles from the solder paste.

Vapor-phase systems are difficult to maintain. Care must be taken to properly maintain vapor-phase systems. The heating elements are a key concern. They must be kept clean of flux residue to avoid overheating the primary fluid. This is accomplished by filtering the primary fluid regularly to remove flux residue.

Vapor-phase fluid is very expensive. The primary fluid used in vapor-phase systems costs approximately $600.00 per gallon.

Vapor-phase systems heat component leads first. With condensation soldering the component leads tend to achieve reflow temperature slightly ahead of the land on the PCB. This causes the solder to wick up the component lead, especially J lead type components. This effect leads to insufficient solder in the solder joint.

8.3 INFRARED REFLOW: LAMP IR AND PANEL IR

The infrared (IR) reflow process is based on the infrared wavelengths of the electromagnetic spectrum, which range from 0.72 microns to 1,000 microns. However, only a small portion of the infrared wavelength is used for reflow soldering. The wavelengths from 0.72 to 1.5 microns are known as near IR; the wavelengths from 1.5 to 5.6 microns are known as middle IR; and the wavelengths from 5.6 microns to 10 microns are known as far IR. The infrared wavelength depends on the temperature of the heat source, in this case an IR lamp or IR panel. All objects emit infrared energy when heated. The rate of radiation, or emissivity, given off by an object increases rapidly as the temperature of that object increases. In an infrared system, heat absorption by the PCA is cumulative. The rate at which the PCA is heated is dependent on the duration of exposure to the infrared energy.

When infrared energy is absorbed by an object it produces heat. At some point during contact with and/or penetration of an object, the infrared energy is converted to heat. The point at which this occurs is dependent on (1) the wavelength of the infrared energy and (2) the absorptivity and/or reflectivity of the material being contacted and/or penetrated. The shorter the wavelength the deeper the penetration will be. As a result near IR will penetrate deeper than middle and far IR.

Materials absorb infrared energy differently. There are four conditions that describe the transmission of infrared energy to an object: reflectivity, opacity, transparency, and translucency. When a material is reflective all of the infrared energy will be reflected from the object; the object is not heated.

When the material is opaque the infrared energy stops at the surface, and only the surface of the object is heated. A material that is transparent allows the infrared energy to pass through the object; the object is not heated. When a material is translucent the infrared energy will penetrate the object to a certain depth and produce heat. Most of the materials used on a typical PCA absorb primarily middle and far IR.

Infrared reflow equipment, based on the emitter type, is separated into two groups: line source emitters (lamp IR) and area source emitters (panel IR). Line source emitters are commonly referred to as lamp IR systems, while area source emitters are commonly referred to as panel IR systems. Panel IR systems are also know as natural (thermal air movement) convection systems because reflow is achieved with a combination of infrared energy and convection. Forced (assisted air movement) convection systems will be discussed as a separate classification. For control, the lamps or panels are arranged in segregated zones.

8.3.1 Lamp IR Systems

Lamp IR systems (as illustrated in Figure 8.8) use line source emitters, which are low-mass heaters. The IR lamps are usually made with tungsten filaments inside a quartz glass tube filled with an inert gas. Most lamp IR systems use an external reflector, which becomes a secondary emitter. Lamp IR/reflector systems emit less near IR and more middle IR and some far IR. Since a gas such as air is transparent to near IR, greater than 90% of the heat generated by either IR lamp or IR lamp/reflector systems comes from infrared energy. Very little heat is produced by conduction and/or convection. Typically, IR lamps will operate above 1,100°C (2,012°F).

Like most other reflow equipment, lamp IR systems have three main sections: preheat, reflow, and cooling. The IR lamps are generally mounted

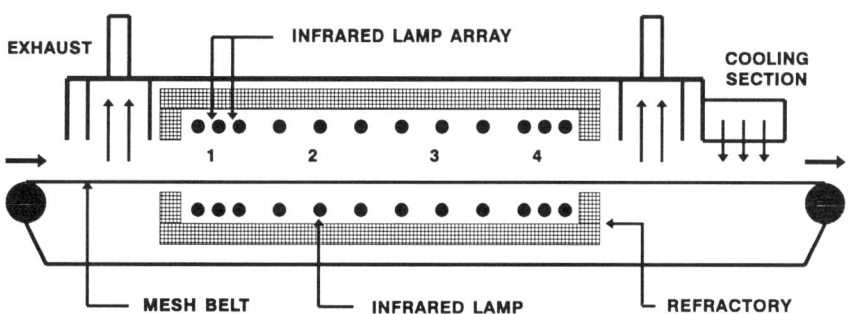

FIGURE 8.8. Infrared Lamp System Concept.

above and below a mesh belt and/or edge-hold conveyor. Preheat sections can contain from four to fourteen zones, with eight zones (four above, four below) being a common set-up. The standard reflow section usually incorporates just two zones (one above, one below). Cooling is primarily provided by one zone of fans (mounted below), but additional cooling can be applied from above. The length of the zone or the number of zones has a direct effect on the transport speed. A longer zone and/or more zones allows a faster transport speed. When infrared energy is used the heat absorbed by an object is cumulative; the heating rate depends upon the amount of exposure to the heat source. Changes to the profile often require the heat source and the transport system to be adjusted.

Lamp IR Attributes

Lamp IR systems have fast response times. Because IR lamps are low-mass heaters they can respond quickly to temperature changes. This can help decrease load sensitivity.

Lamp IR systems can change profiles quickly. Changing from one profile to another can be done quickly because the IR lamp is a low-mass heater and thus does not store heat. Also, because convection does not play a major role, the atmosphere inside the system does not have to be cooled down.

Lamp IR systems offer flexibility. Because IR lamps can achieve high temperatures they may be more useful in some applications than other equipment types.

Lamp IR systems are easy to maintain. Very little maintenance is required, but a good proactive maintenance program is highly recommended.

Lamp IR Concerns

Lamp IR systems emit primarily near IR. Reflow depends on the ability of the object to absorb infrared energy. IR lamps emit mostly near IR; however, most of the materials found on a PCA like to absorb middle and far IR. Color of the object can also be a problem because infrared is not absorbed uniformly across the color spectrum.

Lamp IR systems do not provide equilibrium heating. The object being heated will never (hopefully) reach the temperature of the heating source. This makes profiling complicated. Also, there is a danger that if the mesh belt and/or edge conveyor stops the object being reflowed will be overheated.

Lamp IR systems do not heat uniformly. Typically, the outside edges of the PCB can be 10°C to 20°C (50°F to 68°F) hotter than the center. Also, the low-mass areas on the PCA will heat faster than the high-mass areas, which may cause overheating.

Lamp IR systems have a high source temperature. IR lamps have a very high source temperature, well in excess of 1,000°C.

Lamp IR systems suffer from the "shadow effect." Those areas on a PCA that are not directly exposed to the infrared energy will not heat properly. Components with J leads are particularly difficult, especially if they are tightly spaced.

Lamp IR systems are difficult to profile. Each PCA will require a separate profile because each PCA will absorb the infrared energy differently.

Lamp IR systems can be load sensitive. The IR lamps respond to heat loss by increasing their output. If their response is too great for the mass inside the system overheating can result.

8.3.2 Panel IR Systems

Panel IR systems (as illustrated in Figure 8.9) use area source emitters, which are medium- to high-mass heaters. In general, IR panel heaters (as shown in Figure 8.10) consist of three layers. The face, which can be a secondary emitter, is made from glass, ceramic, or metal. IR panels can be classified as either metallic or nonmetallic, based on the material used for the face. The face, depending on the material used, acts as either a secondary emitter or a transparent window. If the face is made from glass or ceramic it behaves as a window and allows the infrared energy to pass through, in which case the emissivity of the IR panel is based on the emissivity of the primary emitter. If the face is made from metal (aluminum, stainless steel) it absorbs the infrared energy and becomes a secondary emitter, in which case the emissivity of the IR panel is based on the emissivity of the secondary emitter. The primary emitter layer contains a foil or wound resistive element. The insulation layer, or backing, typically contains low-mass refractory insulation.

For proper response and control, location of the thermocouple is critical. A thermocouple is normally placed on the length–width center line of the IR panel. Nonmetallic IR panels place the thermocouple between the face and

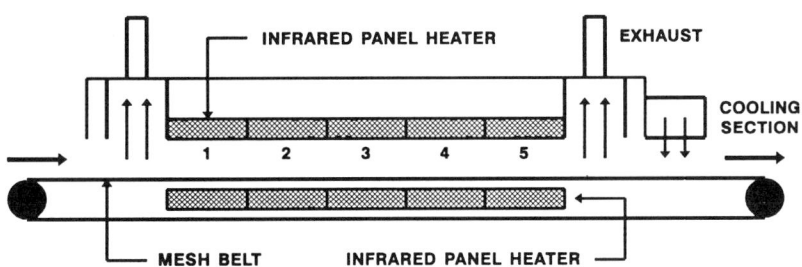

FIGURE 8.9. Infrared Panel System Concept.

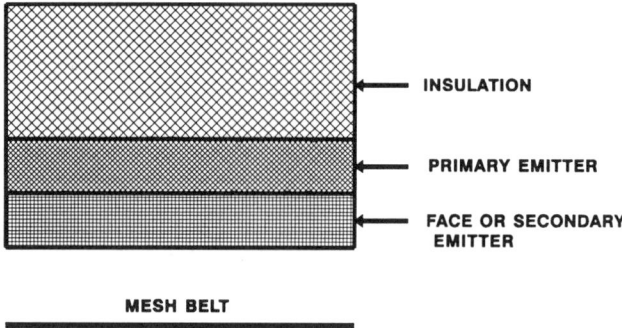

FIGURE 8.10. Example of the Infrared Panel Heater Concept.

primary emitter layers. Metallic IR panels locate the thermocouple on the outside surface of the face.

Heat is produced by a combination of infrared energy and natural convection. IR panels are effective heaters of air (the best IR absorptivity of air is from 5 to 8 microns). The amount of heating created by the infrared energy usually extends from 40% to 50%, and the remaining heat comes from natural convection. Most IR panels operate in a temperature range from approximately 200°C to 450°C (392°F to 842°F).

Like lamp IR reflow equipment, panel IR systems also have three main sections: preheat, reflow, and cooling. The IR panels are generally located above and below a mesh belt and/or edge-hold conveyor. Typical preheat sections contain from four to ten zones, with eight zones (four above, four below) being the routine set-up. Reflow sections are ordinarily comprised of just two zones (one above, one below). The cooling section normally contains one zone mounted from below, but additional cooling can be applied from above. The length of the zone or the number of zones has a direct effect on the transport speed. A longer zone and/or more zones allows a faster transport speed. When infrared energy is used the heat absorbed by an object is cumulative; the heating rate depends upon the amount of exposure to the heat source. Changes to the profile often require the heat source and the transport system to be adjusted. Note the IR panel heaters on the system shown in Figure 8.11.

Panel IR Attributes

Panel IR systems heat with reasonable uniformity. Panel IR systems heat with a combination of radiation and natural convection. The wavelengths emitted by IR panels also heat the surrounding air efficiently. The hot air transfers heat to the PCA more uniformly than just infrared energy alone.

Reflow Soldering and Adhesive Curing 151

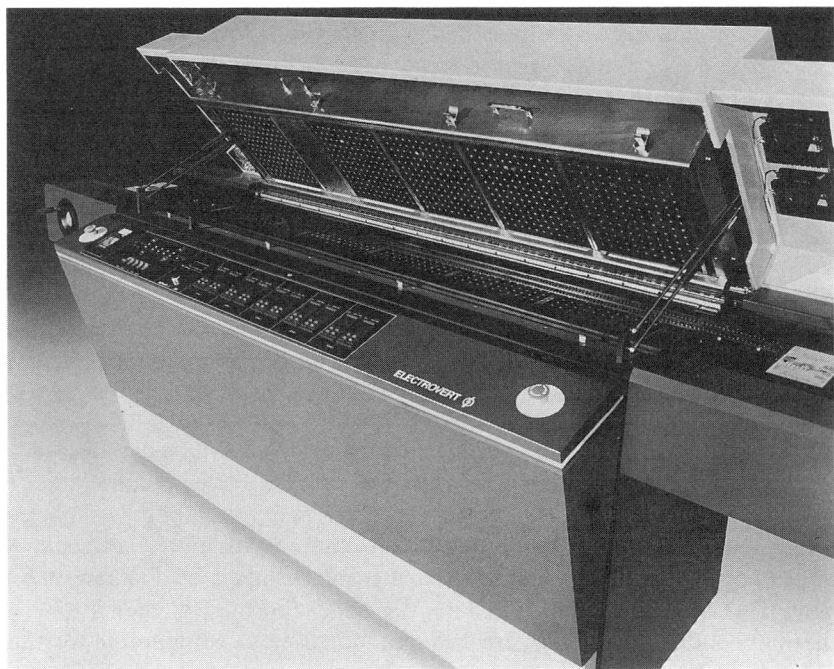

FIGURE 8.11. Example of a Infrared Panel System.

However, the time required to heat objects of different size and mass to the same temperature does depend on their thermal capacity.

Panel IR systems are easy to maintain. Very little maintenance is required, but a good proactive maintenance program is highly recommended.

Panel IR systems are fairly simple to profile. Because the combination of radiation and natural convection heats the object more uniformly, profiling is less critical and easier to achieve than in lamp IR systems.

Panel IR systems emit middle IR and far IR. Most of the materials used on a PCA absorb middle IR and far IR.

Panel IR systems are not color selective. Below 600°C (1,112°F) middle IR and far IR are not color sensitive.

Panel IR systems have a low source temperature. IR panels operate at a temperature between 300°C and 400°C (572°F to 752°F). In contrast, IR lamps operate at over 1,000°C (1,832°F).

Panel IR systems are less load sensitive. IR panels are not as load sensitive as IR lamps. They have a smaller temperature differential to overcome, and they are good thermal reservoirs. However, the IR panels respond to heat loss by increasing their output. If their response is too great for the mass inside the system overheating can result. Metallic secondary emitters are the least load sensitive.

Panel IR Concerns

Panel IR systems transfer energy through radiation. Even though panel IR systems use natural convection, 40% to 50% of the heat transfer still comes from radiant energy. Heat transfer still depends, to a great extent, on the ability of the object being heated to absorb infrared energy and convert it to heat.

Panel IR systems do not provide equilibrium heating. The object being heated will never (hopefully) reach the temperature of the heating source. This makes profiling complicated. Also, there is a danger that if the mesh belt and/or edge conveyor stops the object being reflowed will be overheated.

Panel IR systems can lose heat at their edges. The edges of a metallic IR panel can be cooler than the center, which can result in PCB edges that are underheated. This problem occurs because there is additional surface area on the edges of an IR panel, and this increased surface area dissipates more heat. This occurs more with metallic IR panels because of their excellent thermal conductivity. There are two methods used to compensate for this: Make the IR panels wider than the mesh belt and/or edge-hold conveyor or install edge heaters.

Panel IR systems can overheat at their edges. The edges of a glass and/or ceramic IR panel can be hotter than the center, which can result in PCB edges that are overheated.

Panel IR systems respond slowly to profile changes. This is true when changing to a cooler profile. Because IR panels are good thermal reservoirs it takes time for them to cool.

8.4 FORCED AIR CONVECTION REFLOW

Forced air convection is the most recent development in reflow technology. Heat is transferred to the PCA by low-velocity heated air. Forced air convection is a contact heating method, with little or no heating accomplished by radiation. The rate at which heat is transferred to the object is directly proportional to the difference in temperature between the heated air and the object (PCA).

The air can be heated by several different methods, including IR lamps, IR panels, and conductive elements. The most common and successful forced convection systems use IR panels and/or conductive elements. Air is transparent to near IR, so the air in a lamp IR system must be heated by emissions from a secondary emitter that is emitting middle IR and far IR or by a secondary heating method such as a conductive element. Systems that use IR panels heat the air by forcing it through slots or holes in the panel, where it picks up heat from the panel, and by middle IR and far IR emissions from the panel itself when the air enters the main tunnel of the system. Systems that do not use infrared energy heat the air by bringing it in contact with a conductive element. After the air is heated it is circulated through the main tunnel by a blower. Most forced convection systems recirculate up to 75% of the heated air to increase their heating efficiency. Convection systems must be capable of exact control of air temperature, velocity, and volume.

Based upon the heating methods described above, forced convection systems can be separated into two categories: those that heat the air using radiation, such as IR panels or IR lamps, and those that heat the air with conductive elements.

8.4.1 Infrared Convection

Convection/lamp IR systems are made from modified lamp IR equipment (refer to section 8.3.1 for a complete description). Air is introduced into the heating tunnel through manifolds, inlets in the sheet metal, or small holes in a secondary emitter. The highest absorptivity of air is the wavelengths from 5 microns to 8 microns (middle and far IR). IR lamps are inefficient air heaters since they emit primarily near IR. Some type of secondary emitter or secondary heating method must be present in the system to heat the air efficiently. Convection/lamp IR systems are not really forced convection systems, but rather lamp IR systems that receive some assistance from heated air. Most of the heat transfer is still accomplished with infrared energy.

Convection/panel IR systems are also modifications of panel IR equipment (refer to section 8.3.2 for a complete description). Air is introduced into the heating tunnel by forcing air, usually with a blower, through slots in or between an IR panel, or through small holes in the IR panel. For these systems to be efficient, 50% to 75% of the heated air is recirculated. As noted before, the highest absorptivity of air is the wavelengths from 5 microns to 8 microns (middle and far IR). IR panels are very efficient air heaters since they emit primarily middle and far IR. IR panels also act as thermal reservoirs by storing heat, especially the metallic type. In convection/panel IR systems over 90% of the heat transfer can come from heated air.

Infrared-type forced convection systems have three main sections: preheat, reflow, and cooling. The IR lamps or IR panels are generally located above and below a mesh belt and/or edge-hold conveyor. In some systems the forced air convection is applied from the top only, while other systems apply it from both the top and bottom. Typical preheat sections contain from four to ten zones, with ten zones (five above, five below) being a common

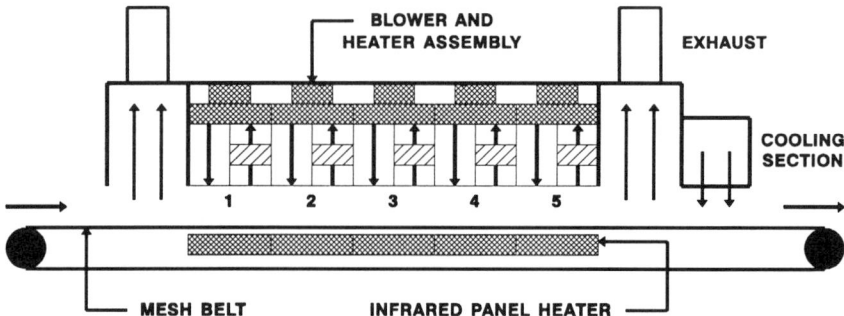

FIGURE 8.12. Forced Convection System Concept.

set-up. Reflow sections are ordinarily comprised of two zones (one above, one below). The cooling section normally contains one zone mounted from below, but additional cooling can be applied from above. The length of the zone or the number of zones has a direct effect on the transport speed. A longer zone and/or more zones allows a faster transport speed. Changes to the profile can often be accomplished by only adjusting the transport speed.

8.4.2 Conductive Convection

The other approach to forced air convection is to heat the air using a conductive element in a closed-loop design. This closed-loop system (shown in Figure 8.12) contains one or more conductive elements, an exhaust port, an intake port, a blower, and one or more measuring instruments, such as thermocouples. Air is continually circulated across the surface of the conductive element by a blower. Heat is transferred from the surface of the conductive element to the air. Fins are often added to the conductive element to increase the heated surface area, which increases the amount of available heat that can be transferred to the air. A baffle is used to disburse the heated air evenly into the main tunnel and onto the PCA. To obtain an efficient heating rate, most systems recycle 50% to 75% of the heated air.

Conductive-type forced convection systems have three main sections: preheat, reflow, and cooling. In some systems the forced convection is applied only from the top side, with IR panels used to supply additional heat from the bottom. Some systems apply forced convection from both the top and bottom. Preheat sections may contain four to sixteen zones, with ten zones (five above, five below) being a common set-up. Reflow sections are ordinarily comprised of two zones (one above, one below). The cooling section usually contains one zone mounted from below, but additional cooling can be applied from above. The length of the zone or the number of zones has a

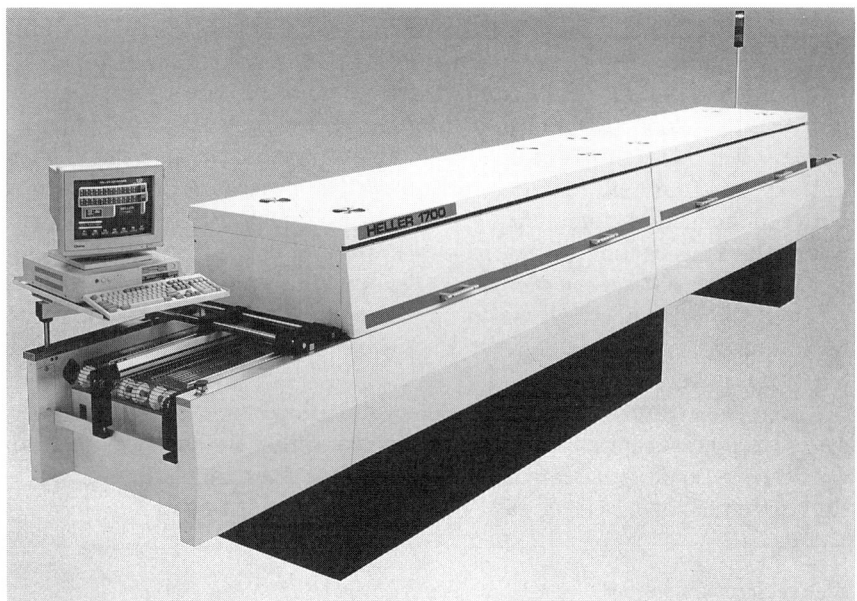

FIGURE 8.13. Example of a Forced Convection System.

direct effect on the transport speed. A longer zone and/or more zones allows a faster transport speed. Changes to the profile can often be accomplished by only adjusting the transport speed. Figure 8.13 shows an example of a forced convection system.

Convection Attributes

Convection systems provide near equilibrium heating. This is especially true with systems that derive greater than 95% of their heat transfer from the heated air. The temperature difference between the PCA and the heated air can be as small as 20°C (68°F).

Convection systems heat uniformly. The hot air transfers heat to the PCA more uniformly than infrared energy. The time required to heat objects of different size and mass to the same temperature does depend on their thermal capacity.

Convection systems are easy to maintain. Very little maintenance is required, but a good proactive maintenance program is highly recommended.

Convection systems are very easy to profile. Because the hot air heats the PCA uniformly, profiling is less critical and much easier to achieve than in infrared systems.

Convection systems are not color sensitive. Hot air does not depend on emissivity and absorptivity to heat the PCA.

Convection systems have a low source temperature. Those systems that do not directly expose the PCA to the heating source, such as conductive element designs, are safer than infrared heated systems because the PCA is not exposed to a high temperature source. The difference in temperature between the heated air and the PCA may be as small as 20°C (68°F) in some cases. The PCA cannot exceed the temperature of the heated air.

Convection systems can change profiles quickly. Cool air can be circulated through the system, which will drop the temperature quickly. Using this technique the system can be reset to a lower temperature much faster than a natural convection system.

Convection systems are not load sensitive. Each zone in a convection system is a good thermal reservoir. Since convection systems provide near equilibrium heating, the introduction of product does not disrupt the equilibrium of the zone, as is the case with systems emitting radiation.

Convection Concerns

Convection systems recycle solder paste and PCB fumes. To make their systems efficient some manufacturers recycle the heated air (50% to 75%) and bring in only a small amount of fresh air (25% to 50%). This method causes vapors from the solder paste and PCB to be recycled back through the system. However, this does not appear to have any detrimental effects on reflow.

Convection systems can lose heat at their edges. Convection systems that use IR panels will suffer from heat loss at the edges because the edges of the IR panel are cooler than the center. This problem occurs because there is more surface area on the edges of an IR panel, and this increased surface area dissipates more heat. This is especially true with metallic IR panels, such as aluminum, because of their excellent thermal conductivity. There are two methods used to compensate for this: Make the IR panels wider than the mesh belt and/or edge-hold conveyor or install edge heaters.

Convection systems must accurately control the air. To operate successfully forced convection systems must have precise control of the air temperature, velocity, direction, and volume.

8.5 CONTROL SYSTEMS

To control their temperatures lamp IR, panel IR, and forced convection systems monitor the PCB temperature, the heater temperature, or the environment temperature. A Proportional-Integral-Derivative (PID) controller is most often used to receive and analyze data from the measurement instrument.

Monitoring the PCB may be the most desirable control method, but it is also the most difficult to implement. An infrared pyrometer could be used to monitor the temperature of the PCB. This complex method, although achievable, is generally not considered practical because of the excessive cost.

A more practical approach is to monitor the heater temperature by placing a thermocouple as close as possible to the heater. This, however, separates the thermocouple from the system environment. This arrangement makes the system less sensitive to load, and the heater may not respond adequately if the load is excessive. Failure to respond will result in a heater that operates at a lower than required temperature.

Monitoring the environment temperature is accomplished by placing a thermocouple near or on the surface of the heater. This format detects changes in the temperature of the system atmosphere. It is important to place the thermocouple as close as possible to the PCB, but without slowing heater response.

8.6 IN-LINE TRANSFER SYSTEMS

Two methods, mesh belt and edge hold, have been developed to transport a PCB. Both methods have merit, depending on how the reflow system is used—either as a stand-alone unit or as part of a continuous assembly line—and whether the PCB has components on one or both sides. The mesh belt conveyor is simple, versatile, and the least expensive technique. Edge-hold conveyors, which support the PCB using two parallel edges, are beneficial when components are on both sides of the PCB and/or when conveyors are used to load and unload the PCB from the reflow system. Some manufacturers offer them together in the same unit, with the edge conveyor mounted above the mesh belt.

Mesh belts, which are made from stainless steel, are available in various widths from approximately 6" to 22". The mesh belt is pulled through the system, usually by a drive sprocket assembly located on the exit end. Proper belt tension is critical. Improper belt tension may cause the belt to jerk, which could move components or cause disturbed solder joints.

Edge-hold conveyors are more complex than mesh belts. Some early designs were complete failures. Fortunately, some good ones are available today. Several different design concepts, some quite simple and some very complex, have evolved. Most are a variation of a standard drive chain. Each design is an attempt at thermal management from two standpoints: do not rob heat from the PCA, and keep the two edges parallel throughout the system. An edge-hold conveyor encounters different temperatures, from ambient to above 400°C (752°F), as it moves through a reflow system. These temperature variations, which occur over a short distance, cause the edge conveyor to expand and contract. The main problem has been with expansion in the reflow zones, which of course have the highest temperatures. If too much expansion occurs the PCB will fall off the edge conveyor. Manufacturers attempt to control this problem in different ways. Some designs attempt to physically contain the expansion using a mechanically strong support arrangement, while others use a low-mass approach with high temperature materials to limit the amount of expansion. At least one design circulates coolant through the edge-hold conveyor to remove heat, and thus limit expansion. The edge-hold conveyor must also compensate for PCB expansion. As noted earlier, the PCB will expand when it is above its Tg temperature. The edge-hold conveyor must allow this expansion to occur without putting pressure on the PCB. And, finally, the edge-hold conveyor must not steal heat from the PCB.

As noted previously, some manufacturers offer a mesh belt/edge-hold conveyor combination with the edge-hold conveyor mounted above the mesh belt (as shown in Figure 8.14). This combination offers maximum flexibility to the user. It also provides an important safety feature when the edge-hold

FIGURE 8.14. Example of a Mesh Belt/Edge Hold Conveyor.

conveyor is used. If a PCB drops off the edge-hold conveyor it will land on the mesh belt, which will transport the PCB safely out of the reflow system. A PCB can drop off because of a poorly designed and/or maintained edge-hold conveyor or a weakening of the PCB due to reflow temperatures that heat the PCB well above the Tg temperature. The mesh belt/edge-hold conveyor combination can be well worth the additional cost because of increased flexibility and safety.

A closed-loop system that can monitor mesh belt and/or edge-hold conveyor speed is a valuable option. Some of the lower priced reflow equipment may not offer this as a standard feature, so be sure to check with the manufacturer. Another significant factor is ease of access. Many systems are hinged to allow easy access to the edge-hold conveyor and mesh belt in the preheat and reflow zones. For maintenance purposes, this can be very helpful.

8.7 CONTROLLED ATMOSPHERES

The atmosphere inside the reflow system plays an important part in the reflow process. As noted earlier, the vapor-phase process provides a controlled (inert) atmosphere during reflow. Most other reflow systems (lamp IR, panel IR, convection) use an uncontrolled air atmosphere; however, these systems can be set up to use a controlled atmosphere.

A controlled atmosphere can assist the reflow process by enhancing solderability, improving flux efficiency, preventing flux charring, and limiting discoloration of the PCB and solder mask. Controlled atmospheres can be of two types: protective or reactive. A protective (inert) atmosphere does not react with the materials being soldered. A reactive atmosphere will cause a reaction with the materials being soldered. The purpose of the protective atmosphere is to prevent oxidation of the metal surfaces during the reflow process. On the other hand, a reactive atmosphere is intended to aid in the removal of oxidation during the reflow process. Protective atmospheres may aid the reflow process, depending on the conditions. Reactive atmospheres on the other hand have yet to prove effective in aiding the reflow process.

Common controlled atmospheres include nitrogen, hydrogen, ammonia, nitrogen/hydrogen blends, and nitrogen/methanol blends. These gases may be inert, active, or reducing. Nitrogen is the most common controlled atmosphere and is used to provide an inert environment inside the reflow system.

8.8 TIME/TEMPERATURE PROFILING

Printed circuit board time/temperature profiling is extremely important, yet it is often misunderstood and improperly used. Many important and possibly expensive decisions (such as determining which reflow system to purchase,

160   Applied Surface Mount Assembly

process development, and process verification) are made based on the results of profiling. The profiling information presented here can be applied to lamp IR, panel IR, forced convection, and vapor-phase reflow systems.

Once a time/temperature profile is developed it must be applied to the reflow system. This is done using profiling equipment and operator experience. Several profiling systems are available. They range from simple recorders to complex PC-based systems that can do simulation and process monitoring, as well as provide thermocouple data (as shown in Figure 8.15).

8.8.1 Thermocouples

Temperature profiling requires rapid, reliable measurement. This is accomplished by the use of insulated, exposed junction thermocouples. Thermocouples are manufactured using solid wires covered with a glass braid or other high temperature insulation. A common wire diameter for PCB temperature profiling is 1.59mm (0.0625") (30 AWG). Two common thermocouples used for profiling are Type J and Type K. Type J thermocouples, iron versus

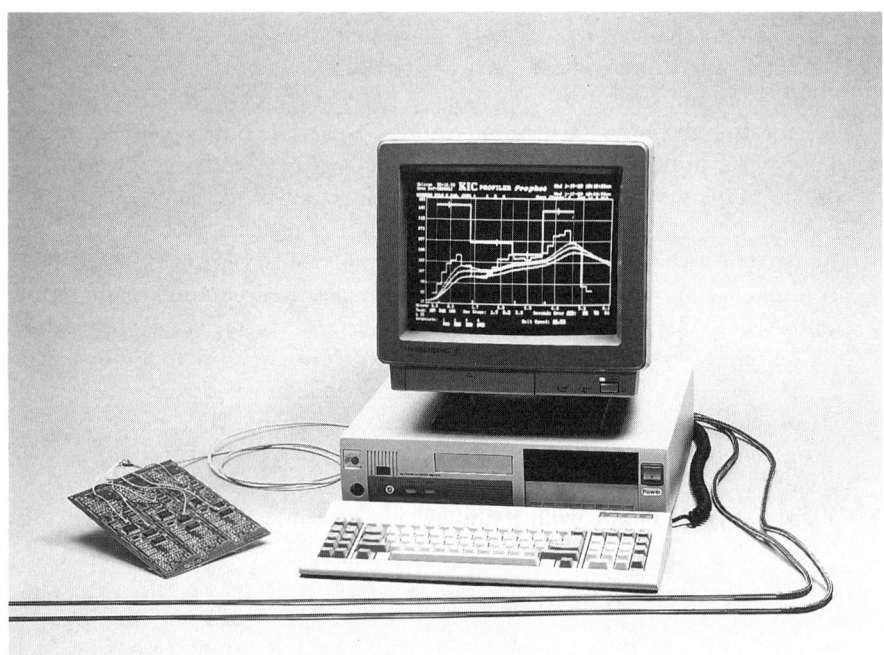

FIGURE 8.15.   An Advanced Time/Temperature Profiling System.

copper/nickel, are rated to 316°C (537°F) for a 30 AWG wire with a tolerance of ±2.2°C from 0° to 277°C (531°F). Above 277°C (531°F) the tolerance is ±0.75%. Type K thermocouples, nickel versus nickel/aluminum, are rated to 760°C for a 30 AWG wire with a tolerance of ±2.2°C from 0° to 277°C (531°F). Above 277°C (531°F) the tolerance is ±0.75%. The temperature range will also depend on the type of insulation used.

8.8.2 Thermocouple Junction

Three methods are used to join thermocouple wire together: welding, soldering, and twisting. To operate, a thermocouple generates a voltage at the junction of the two wires. The voltage will change as the temperature changes. If more than one junction is made, a voltage will be generated at each one and the output voltage will be an average of all of the junctions. Also note that junction reliability declines with use. Each junction should be inspected before being used and replaced when necessary.

Welding, if done correctly, can produce the best thermocouple junction; however, it is the most difficult and expensive technique. It requires expensive welding equipment and should be done in an inert atmosphere to prevent oxidation of the wire. Commercially supplied thermocouples use this method.

Soldering the wires together requires a high-temperature solder (95Sn/5Pb) so the junction will not separate during reflow. Soldering is the least recommended of the three methods.

Twisting the two wires together is the simplest technique. Place a small object, such as a pin, between the two wires. Then, twist the wires tightly together with a minimum number of turns. Remove the pin and trim off the excess wire.

8.8.3 Thermocouple Attachment

The thermocouple junction must be attached to the PCB in such a way that it is thermally conductive and electrically insulated. Merely laying the thermocouple junction on the PCB and taping it in place is not acceptable. The entire thermocouple junction must make contact with the PCB or component lead to ensure a good conductive path.

Two methods, soldering or gluing, are used to attach the thermocouple junction. Soldering is done using a solder alloy with a high enough melting temperature that the solder will not melt during the reflow process. The thermocouple junction is soldered to a land, pad, or component lead. The solder joint should be as small as possible, but completely cover the thermocouple junction. The small solder joint will allow proper heating of the thermocouple junction. Lightly clean the solder joint to remove any flux.

To attach the thermocouple junction with adhesive, select an easy-to-use adhesive that has good thermal conductivity (minimum 1.0 Watts/C Sec), and is also electrically insulating. The thermocouple junction is glued to a land, pad, or component lead. Hold the thermocouple junction in position and cover with adhesive. Apply a small amount of adhesive, but completely cover the thermocouple junction. A minimum amount of adhesive will ensure proper heating of the thermocouple junction. Support the thermocouple junction until the adhesive has cured properly.

The thermocouple wire should be attached securely to the PCB near the point of attachment, and then approximately every one to two inches. A thin wire or high-temperature tape can be used. A strain relief may also be provided for the thermocouple wire, just after the point of attachment. Finally, any bare thermocouple wire should be covered with an insulating adhesive to prevent oxidation. Heat from the reflow process will encourage oxidation, and any exposed wire will rapidly deteriorate.

8.8.4 Thermocouple Placement

Thermocouples should be placed at various locations on the PCB to give an overall indication of the temperature across the PCB. As a minimum, the two density extremes (low and high) on a PCB should be studied. In other words, if there is an area with few or no components this would be the low-density extreme. The high-density extreme would be an area with a large number of difficult-to-reflow components, such as the PLCC. The low-density area will, of course, require less heat, while the high-density area will require more heat to achieve the same temperature. The thermocouples should be used to confirm the low-density area is not overheated and the high-density area receives the correct amount of heat. In low-density areas it may be best to attach the thermocouple to the PCB to ensure that it is not overheated. In high-density areas or when evaluating difficult to reflow components the thermocouple should be attached to the component lead to confirm that the lead is heating correctly.

8.9 ADHESIVE CURING

After adhesive is applied to a PCB a component is placed in the adhesive. The next step is to cure the adhesive so the component will be held in place during the wave soldering process. Heat is the primary method used to cure adhesives. Acrylics can also be cured using ultra violet (UV) light, this is usually applied in conjunction with heat because the UV will only cure the exposed adhesive that it can make contact with.

Heat curing can be accomplished with a convection or infrared system. Vapor-phase is not recommended for curing. Some convection and infrared systems can also be fitted with a UV zone at the beginning of the system. Temperature is more important than time for achieving high bond strength. Another important consideration is the type of solder mask used on the PCB. Bond strength will vary between solder-mask materials.

Recommended time/temperature curing profiles should be obtained from the adhesive supplier. See Chapter 5 for adhesive information.

REFERENCES

1. Avramescu, Sabi. "The Evolution in Reflow Soldering Systems." *Circuits Manufacturing*, November 1989, pp. 28–33.
2. Baker, James. "Reflow Technology—A User's Perspective." Proceedings, EXPO SMT '89, September 1990, pp. 113–118.
3. Beiser, Arthur. *Physics*. Menlo Park, CA: Cummings Publishing Company, 1973.
4. Bergenthal, Jim. "Reflow Soldering Process Considerations for Surface Mount Application." KEMET Corporation, Application Bulletin, January 1989.
5. Cox, Norm. "Combining Radiation and Convection." *Circuits Assembly*, April 1991, pp. 43–46.
6. Dow, Steve. "The Use of Zone Segregated Full Convective Heat Transfer in Mass Reflow Soldering." Proceedings, Surface Mount '90, August 1990, pp. 105–115.
7. Eck, Dave. "Infrared Heaters—Determining Output and Choosing the Right One," Internal Solar Products Report, 1990.
8. Flattery, David. "An Overview of the Selection of and use of Convection/Infrared Reflow Equipment." Proceedings, EXPO SMT '89, September 1990, pp. 307–311.
9. Glynn, Michael R. "Facilities Considerations in the Selection of Solder Reflow Equipment." Proceedings, Surface Mount '90, August 1990, pp. 116–121.
10. ———. "Full Spectrum IR Reflow Through Energy Management." *Printed Circuit Assembly*, February 1989.
11. Hutchins, Charles C. "Make the Most of Your Reflow Environment." *Circuits Manufacturing*, 1990.
12. ———. "SMT/FPT Soldering Problems and Solutions." *Surface Mount Technology*, July 1990, pp. 37–40.
13. ———and King, Scott. "The Surface Mount Reflow Process." *Printed Circuit Assembly*, April 1987, pp. 21–24.
14. IPC-SM-786. "Recommended Procedures for Handling of Moisture-Sensitive Plastic IC Packages." IPC, Lincolnwood, IL, April 1991.
15. Johnson, Colin C. and Kevra, Joseph. *Solder Paste Technology*. Blue Ridge Summit, PA:Tab Books, Inc., 1989.
16. Kasturi, Sangita. "Forced Convection: A User's View." *Circuits Manufacturing*, September 1990, pp. 68–74.
17. Kazmierowicz, Philip C. "Profiling Your Solder Reflow Oven in Three Passes or Less." *Surface Mount Technology*, February 1990.

18. Martel, Michael L. "Forced Convection: The Dark Horse." *Circuits Assembly*, February 1989, pp. 27–40.
19. Prasad, Ray and Aspandiar, Raiyomand. "VPS and IR Soldering for SMAs." *Printed Circuit Assembly*, February 1988, pp. 29–37.
20. Samsami, Darius. "SMT Reflow: Facing the Challenges." *Electronic Packaging and Production.*" January 1991, pp. 65–67.
21. Uehling, Trent and Glynn, Michael. "Know thy Thermocouple." *Circuits Manufacturing*, September 1989.
22. Zarrow, Philip. "IR Reflow Soldering Equipment." Proceedings, EXPO SMT '89, September 1990, pp. 79–81.
23. ———. "IR Reflow Soldering Systems and Steps." *Circuits Manufacturing*, February 1990, pp. 26–38.
24. ———. "Optimizing IR Reflow." *Electronic Packaging and Production.* " November 1990, pp. 32–40.
25. ———. "Selection Criteria for Infrared Soldering Systems." Proceedings, NEPCON West, February 1990, pp. 841–850.

9
Wave Soldering

GLOSSARY

Bidirectional Waveform   A waveform that allows the solder to flow away from the nozzle in two opposing directions.

Dross   A waste by-product that forms on the surface of molten solder. It consists primarily of tin oxide, lead oxide, and flux residues.

Dual Wave System   A solder module developed for wave soldering surface mount components. It consists of a turbulent waveform and a laminar waveform.

Dwell Time   The amount of time any one part of the PCA is in contact with the solder wave.

Entry Angle   The angle at which the PCA makes contact with the solder wave.

Flux Module   The unit used to apply flux to the PCA. It usually consists of a fluxer, an air knife, a flux storage container, and a flux density controller.

Laminar Waveform   A waveform that directs the solder to flow primarily in one direction. Also referred to as a unidirectional or smooth wave.

Preheat Module   The unit used to heat the PCA prior to wave contact. It consists of element or infrared (IR) heaters located below, and possibly above, the PCA. Insulated covers and measuring instruments are available options.

Profile   The relationship of time versus temperature during the wave soldering process.

Solder Module   The unit used to apply solder to the PCA. It consists of a solder pot, heating elements, pumps, and nozzles. It may contain one or two waveforms.

Vibrating Wave System   A solder module developed for wave soldering surface mount components. It consists of a single vibrating laminar wave.

Volatile Organic Compound (VOC)   In general, an organic compound that is readily vaporizable in ambient air. The exact definition varies depending on local government regulations.

9.0 INTRODUCTION

Wave soldering has an important, but limited, role in surface mount assembly. This process, originally developed for soldering pin through-hole (PTH) components, should be used primarily for soldering surface mount passive components. Some modifications have been made to the original process to allow soldering of surface mount components. The basic process is made up of five constituents: PCB handling, flux application, preheat, soldering, and cool down. The central element in surface mount wave soldering is wave geometry. Designing for assembly plays a vital role in this process, probably even more so than with reflow soldering. Prior to soldering, the components are attached to the PCB using an adhesive. Information on adhesives, application methods, and curing can be found in Chapters 5, 6, and 8 respectively.

This chapter will examine time/temperature profiles, designing for assembly, flux application, preheat, soldering application and PCB handling.

9.1 TIME/TEMPERATURE PROFILES

A time/temperature wave soldering profile should be developed before implementing the process. Several factors influence a wave soldering profile. Each one should be clearly defined and understood. The wave soldering profile can be described in three phases: preheat, wave contact, and cooldown. Preheat is needed to heat the PCA at a uniform and controlled rate, evaporate flux solvents, and activate the flux. Solder joints are formed by contact with a molten solder wave. Transport speed controls dwell time in the solder wave. Cool down provides the proper cool-down rate for the solder joint so the proper grain structure is formed.

Examples of generic profiles can be found in Figures 9.1, 9.2, and 9.3. Figure 9.1 defines a time/temperature profile for PTH soldering, shown so that the other two profiles can be compared to it, and the difference between PTH soldering and SMT soldering can be noted. Figure 9.2 is a time/temperature profile for a dual wave SMT soldering system, while Figure 9.3 is a time/temperature profile for a single wave SMT soldering system. Note that a major difference between the PTH and SMT time/temperature profile is the maximum preheat temperature. SMT wave soldering requires a smaller temperature delta between preheat and the solder wave to eliminate component thermal shock.

During the preheat phase the PCB should be heated at a rate between 0.5°C/second (0.9°F/second) and 2°C/second (3.6°F/second) until the surface mount components reach a temperature that is within 100°C (212°F) of the molten solder temperature. Note that there is a slight decrease in temperature after the PCA exits the preheat phase and before contact with the solder

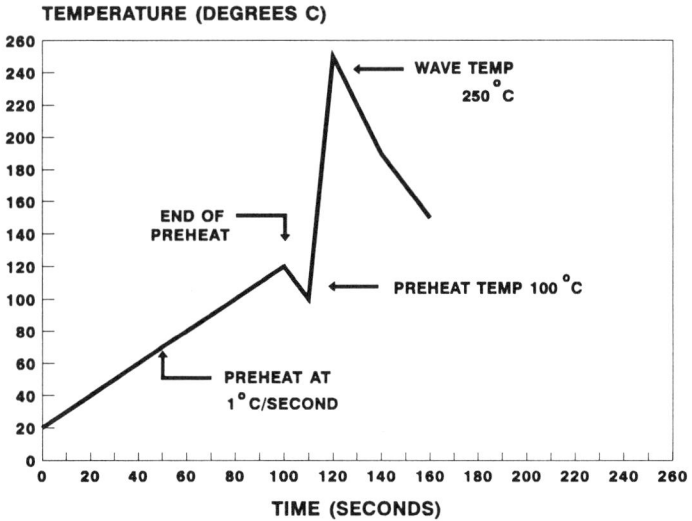

FIGURE 9.1. A Common Standard Time/Temperature Profile.

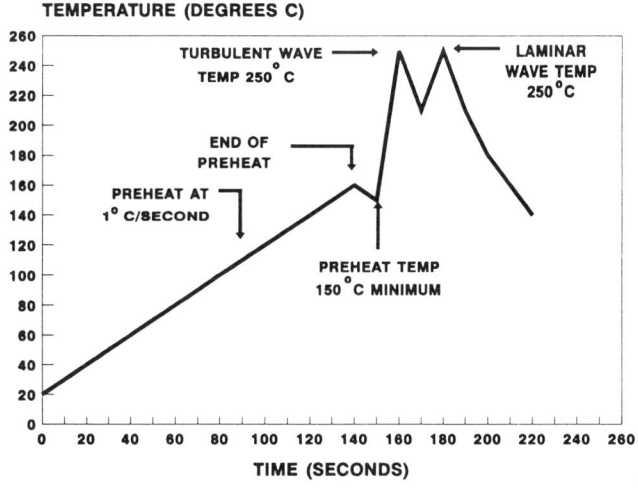

FIGURE 9.2. A Common Dual SMT Wave Time/Temperature Profile.

SINGLE SMT WAVE PROFILE

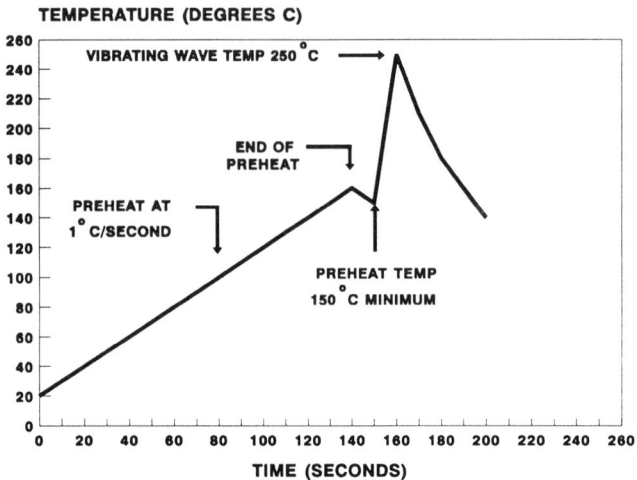

FIGURE 9.3. A Common Single SMT Wave Time/Temperature Profile.

wave. To properly maintain the 100°C (212°F) temperature delta it may be necessary, depending on the equipment used, to preheat the PCA slightly above the required preheat temperature to compensate for the heat loss. Molten solder temperatures range from a minimum of 240°C (464°F) to a maximum of 260°C (500°F), but the lower temperature of 240°C (464°F) is preferred because this would also decrease the preheat temperature required to achieve the 100°C (212°F) temperature delta. For information on profiling equipment and methods refer to section 8.8.

9.1.1 Printed Circuit Boards

As noted in Chapter 8, printed circuit boards can be damaged by excessive exposure to heat. For the wave soldering process it is important to understand the glass transition temperature (Tg) of the PCB for two reasons. Above the Tg the PCB will soften and lose its rigidity. This is a problem for the PCB handling system. If the PCB loses too much rigidity it will fall off the handling system. Also, above the Tg the expansion rate of the PCB increases dramatically, especially in the unsupported Z axis. This expansion can damage the PCB, particularly the plated through-holes. The time above Tg should be limited to around 120 seconds.

Thermal shock occurs because of the difference in temperature between the top side of the PCB and the bottom side of the PCB during wave contact.

This temperature difference can severely warp the PCB, which can cause major problems during wave contact. Preheating reduces or eliminates thermal shock. Preheat temperatures vary depending on the number of layers and the PCB thickness. A double-sided PCB should be preheated to a temperature range of 100°C to 110°C (212°F to 230°F). A multilayer PCB up to approximately 1.5mm (0.060") thick should be preheated to a temperature range of 110°C to 120°C (230°F to 248°F). A multilayer PCB up to approximately 2.5mm (0.100") thick should be preheated to a temperature range of 120°C to 130°C (248°F to 266°F). Temperature is measured at preheat exit, on the top-side laminate. These temperatures are recommended guidelines. Some PCBs may require more or less preheat. If wave soldering surface mount components, additional preheat may be required (refer to section 9.1.3). Preheating is not a substitute for baking the PCB to remove absorbed moisture.

The surface finish on the lands and plated through-holes will have a significant impact on solderability. For more information refer to Chapters 3 and 5.

9.1.2 Flux

Flux is affected by two factors during the wave soldering process: temperature and time. Most rosin-type fluxes become active at approximately 110°C (230°F), while organic acid fluxes are usually active at room temperature. However, an elevated temperature does improve the fluxing action of an organic acid flux. The flux also needs to be active long enough to properly remove oxidation and contamination. An acceptable activation time for most fluxes is 30 seconds minimum and 90 seconds maximum. It is important to avoid overheating and burning or charring the flux.

Heat is also necessary to remove volatiles (thinner) from the flux. Water-based fluxes may require more preheat than solvent-based fluxes. Most fluxes are classified as a volatile organic compound. Some local government regulations are placing severe restrictions on the use of VOC classified materials.

9.1.3 Components

Components are very sensitive to the wave soldering process. The only components suitable for wave soldering are rectangular and cylindrical ceramic and glass resistors, capacitors, and diodes. Molded components, such as the small outline (SOT, SOIC, SOLIC, SOM, SOP) package, can be wave soldered, but there is a concern about flux seeping into the package through the lead frame to plastic interface. This is a potential reliability problem with all molded packages. Plastic leaded chip carriers (PLCC) have another problem, in addition to the flux issue. The wave soldering process applies too much solder to the J lead; this will affect lead compliance. If the lead becomes too stiff more pressure is applied to the solder joint because the lead can not expand properly when subjected to operating temperatures.

Components can be damaged by the incorrect application of heat during the wave soldering process. All components have a limit as to the maximum temperature and heating rate to which they can be exposed. Most surface mount components can withstand a 260°C (500°F) temperature for ten seconds, which provides an acceptable margin since contact with the solder wave typically lasts three to four seconds. Thermal shock, caused by a rapid increase in temperature during preheat or contact with the solder wave, has been a major concern, especially with capacitors. During preheat, components should not be exposed to more than a 2°C/second (3.6°F/second) heating rate. To prevent thermal shock during contact with the solder wave, it is important to preheat the components to within 100°C (212°F) of the molten solder temperature.

Component lead finish is an important factor that will affect solderability and reliability. For information on this subject refer to Chapters 2 and 5.

9.2 COMPONENT LAYOUT

As discussed previously in Chapter 4, the layout, or orientation, of the components in wave soldering is critical. Proper orientation helps overcome two problems: flux entrapment and shadowing. Rectangular and cylindrical components must be oriented perpendicular to the direction of travel through the wave soldering system. Leaded components, such as an SOT or SOIC, should be oriented parallel to the direction of travel through the wave soldering system (the leads are actually oriented perpendicular to the direction of travel). Proper component orientation is illustrated in Figure 9.4.

Gas bubbles are generated when the flux contacts the solder wave. These gas bubbles adhere to the component terminations and leads. The trapped gas bubbles interfere with the solder wave, which can prevent a solder joint from forming. Gas bubble formation can be reduced with proper preheating. Flux entrapment is not a problem with through-hole wave soldering because the gas bubbles have a natural escape path, as shown in Figure 9.5. However, as shown by Figure 9.6, a surface mount component provides no natural escape path for the gas bubbles. Two constituents are required to displace the trapped gas bubbles: proper component orientation and a turbulent or agitated solder wave (discussed later in this chapter). Proper component orientation permits the solder wave to dislodge the gas bubbles, allowing the solder joint to form.

Surface mount components provide termination and lead geometries that are difficult to wet with solder. In plated through-hole soldering, the curved surfaces of the hole and lead help the wetting action of the solder. Plated through-hole soldering also benefits from capillary action pulling the solder

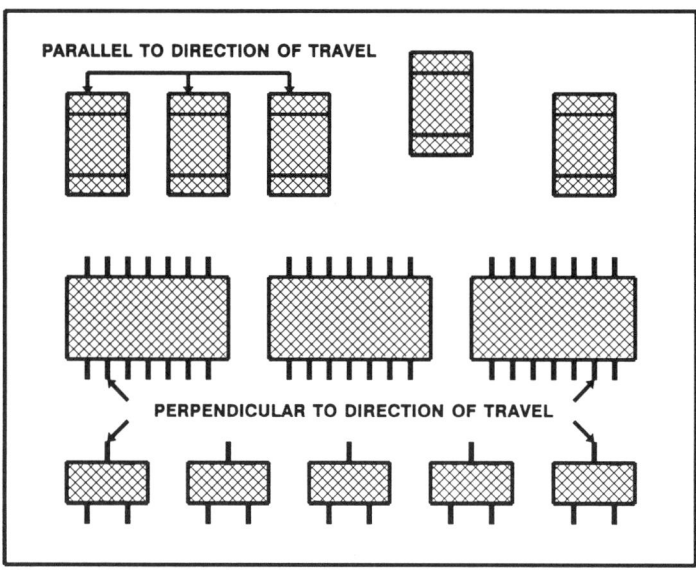

FIGURE 9.4. Preferred Component Orientation for Wave Soldering.

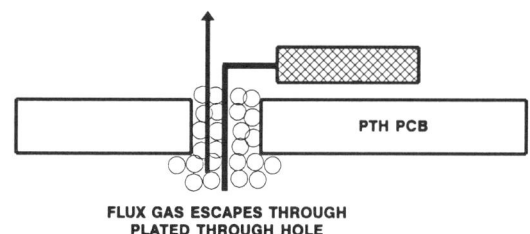

FIGURE 9.5. Flux Escape Path for Insertion Mounted Components.

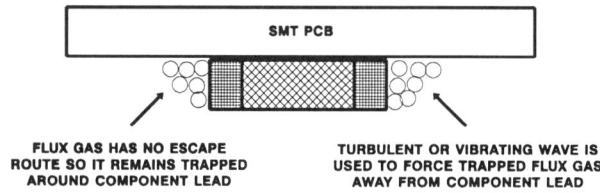

FIGURE 9.6. Flux Trapped Around a Surface Mount Component.

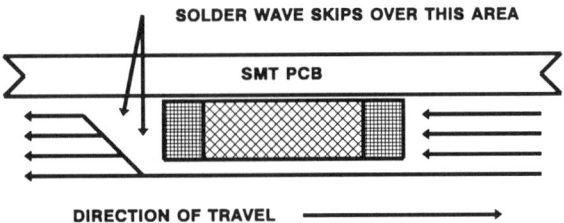

FIGURE 9.7. Solder Skips Due to Incorrect Component Orientation.

up the hole. In surface mount soldering, the flat surfaces of the PCB land and component termination cause a standard laminar solder wave to contact these surfaces tangentially, which does not promote solder joint formation. If the component is oriented parallel to the direction of travel rather than perpendicular, the trailing termination will be skipped over completely by the solder wave, as indicated by Figure 9.7.

9.3 FLUX APPLICATION

Three systems are currently available to apply flux to a PCA: foam fluxers, wave fluxers, and spray fluxers. The most common and popular approach is the foam fluxer. Spray fluxers are becoming popular for the application of low solid fluxes (LSF).

9.3.1 Flux Module

A foam fluxer, as shown in Figure 9.8, is a fairly simple apparatus used for applying flux to short leads. It consists of an open container, or reservoir, for the flux, with a porous stone placed on the bottom and a nozzle located on the top. The container is filled with flux, submerging the stone. Foam is generated by pumping air through the porous stone. Air pressure then forces the foam to the top of the nozzle, where a stable foam head of fine flux bubbles is created the full width of the fluxer. Excess flux cascades down the outside of the nozzle and returns to the reservoir. Maintaining a stable foam head can be a challenge. Keeping the stone clean is critical. A stable, dry, oil-free air supply is also essential. The flux reservoir should be covered when not in use to prevent thinner evaporation. When not adjusted correctly, foam fluxers can flood the top side of the PCA with flux. Most fluxes for foam fluxers benefit from additives that promote foaming. An LSF may be difficult to foam because of the low-solids content and the absence of foaming agents. An air knife is usually placed after the fluxer to push the flux into the plated through-holes and remove excess flux.

FIGURE 9.8.   Example of a Foam Fluxer.

A wave fluxer is also a fairly simple device. It consists of a container, or reservoir, for the flux, a small pump, and a nozzle. The flux is pumped up and over the nozzle, creating a wave of flux. Wave fluxers tend to apply a heavy coating of flux. Brushes or an air knife are usually located after the fluxer to remove excess flux. This method is useful for long leads or when a heavy coating of flux is required. Wave fluxers, if not properly adjusted, have a tendency to flood the top side of the PCA. The wave surface must be very level to ensure an even coating of flux on the PCA. The reservoir should be covered when not in use to prevent thinner evaporation.

Spray fluxers have become popular as a method for applying LSF. Three types of spray fluxers have been developed: high-velocity spray, rotating drum spray, and ultrasonic spray. If flammable materials are used fire prevention is a concern. Most spray fluxers are located in external modules because of fire and overspray concerns.

The high-velocity spray fluxer consists of a spray-type nozzle or series of nozzles, through which flux is forced under pressure. The spray applies a thin coating of flux to the bottom side of the PCA. High-velocity spray fluxers can suffer from nonuniform flux application and clogged nozzles. These fluxers usually draw flux and thinner directly from the suppliers' containers, which makes them a closed system.

Rotating drum fluxers use a rotating mesh cylinder, or drum, that is partly submerged in a container of flux. An air knife is mounted inside the cylinder, on its center line, positioned to spray air upward. As the cylinder rotates flux collects on the mesh. The air knife sprays air through the mesh, creating a fine mist of flux that is sprayed on the bottom of the PCA. This method is somewhat unique because the amount of flux applied to the bottom of the PCA can be varied by using a different mesh on the cylinder, changing the revolution speed of the cylinder and by altering the air pressure.

An ultrasonic spray fluxer generates a flux mist by exposing liquid flux to high-frequency sound waves inside a chamber. The flux mist is forced upwards and deposited on the bottom of the PCA by a low-velocity current of air. The amount of flux deposited on the bottom of the PCA is controlled by the volume of flux flowing into the chamber, while the sound-wave amplitude controls the size of the individual flux beads. This is also a closed system. The ultrasonic spray fluxer was developed specifically for applying an LSF.

9.3.2 Flux Density Control

An automatic flux density control system is a very valuable asset. A flux density control system typically consists of a density monitor and three reservoirs: the main flux container, a reserve flux container, and a reserve thinner container. All three containers are filled with the proper materials. The flux density is usually monitored in two locations: the main flux container and in or near the flux module. If the density is incorrect, flux or thinner is drawn from the proper container and added to the main flux container until the proper density is achieved.

It is also important to change the flux on a regular schedule. The recommended interval should be obtained from the flux supplier.

9.4 PREHEAT

Preheating is a very important part of the wave soldering process. It has four objectives: remove solvents by evaporation, activate the flux, eliminate thermal shock to the PCB and components, and enhance soldering speed.

9.4.1 Preheat Objectives

Issues relating to time and temperature were discussed in section 9.1. Preheating improves soldering speed by bringing the surfaces to be soldered to their wetting temperature more quickly. If more heat is supplied by preheat then less heat needs to be supplied by the molten solder. In other words, proper preheat reduces dwell time.

9.4.2 Preheat Modules

Preheat modules are available in many formats, but they all contain the same basic characteristics. The most important of these is the heating unit. Heating is achieved by convection or radiant energy, or a combination of both. Convection heating is provided by conductive element heaters, installed as rods or panels. Radiant heating is provided by infrared (IR) lamps or panels. Conductive and radiant heaters are discussed in detail in Chapter 8. The heaters are positioned to cover the full width of the system. Heater length varies from a minimum of approximately 2' to a maximum of 6' to 8'. Heaters are always located on the bottom. Additional heaters can be provided on the top. If heaters are not added to the top, an insulated cover, or tunnel, is typically used to help contain the heat from the bottom heaters. Most heating modules contain two or three zones across their width. The zones are turned on or off depending on the width of the PCB. A roll-out drip pan is located under the heaters to contain excess flux and solvent that drips from the bottom of the PCB. Aluminum foil is usually placed in the bottom of the drip pan to aid clean up. It also acts as a reflective surface that redirects heat toward the PCB. If the heating elements are exposed, a glass plate is usually placed over them to prevent flux from dripping onto the elements. The exit end of the preheat module is placed as close as possible to the solder module to limit PCB temperature loss before wave contact. Part of a preheat module is presented in Figure 9.9.

9.4.3 Temperature Monitoring

Two simple methods are available to monitor preheat temperature. A heat sensitive label, which changes color, is placed on the top-side laminate. The color of the label will change depending on the maximum preheat temperature. The other method is to use a small chip from a heat sensitive stick. A small chip is cut from the stick and placed on the top-side laminate. The chip melts after achieving a certain temperature. Both the labels and the sticks are available in an assortment of temperature ranges. These methods are "go/no-go" indicators, they do not provide the user with the actual preheat temperature.

To obtain more detailed information, preheat temperature can be measured using a profiling system and thermocouples, or by using an integrated pyrometer. The pyrometer has an advantage because every PCB can be measured in real time. Profiling systems and thermocouples were discussed in detail in Chapter 8. A pyrometer, which measures PCB emissivity, is placed

176   Applied Surface Mount Assembly

FIGURE 9.9.   Example of a Preheat Section.

near the end of the preheat module above the PCB. Most pyrometer set-ups are adjustable, either manually or automatically, across the width of the PCB. The pyrometer will determine the top-side PCB temperature, and then alert the operator if the temperature is not correct.

9.5 SOLDER APPLICATION

Contact with the solder wave achieves the objective of forming solder joints on the PCA. Everything that has been discussed up to this point—component layout, flux application, and preheat—prepares the PCA for wave contact. Wave contact supplies two important elements: heat transfer and material transfer. Surface mount wave soldering has forced the development of new wave geometries that can dislodge trapped flux gas bubbles and properly form solder joints at undesirable solder wave contact angles.

9.5.1 Wave Soldering Basics

There are seven important variables associated with wave soldering: solder alloy, solder temperature, entry/exit speed, dwell time, wave geometry, contact length, and entry angle. Solder alloys and temperature are discussed in Chapter 5. Entry and exit speed is controlled by the PCB transport system. Dwell times of one to two seconds support good solder joint formation. Wave geometry varies according to the equipment supplier.

Wave geometry has experienced continued development and improvement over the past ten years. The solder module consists of a heating system, a pumping system, a wave generating apparatus, and a solder pot. Solder is pumped from the bottom of the solder pot by an impeller or propeller, which prevents dross from being circulated through the wave. Wave contact can be divided into four events: entry, heat transfer, material transfer, and exit. The PCB enters the wave, flux is removed by a scrubbing action, and heat transfer begins. The PCB continues into the wave, heat transfer is complete, and material transfer begins. The plated through-holes fill with solder, and the solder joints form. The PCB exits the wave, peel-back forces pull unwanted solder back into the solder pot, eliminating bridging and icicling. A laminar wave is designed so that the surface speed of the molten solder matches the PCA exit speed.

The original waveform is referred to as a bidirectional wave, since the solder is pumped upward and then allowed to cascade out in two directions as shown in Figure 9.10. In recent years the bidirectional wave has been modified into a unidirectional wave, commonly referred to as a laminar or smooth wave. The laminar waveform, also pictured in Figure 9.10, was devised to help reduce icicling and bridging. A laminar waveform diverts part of the cascading solder back toward the center of the wave, which creates a solder flow in two opposing directions. When these opposing solder flows collide they create a stationary zone of molten solder. The stationary zone

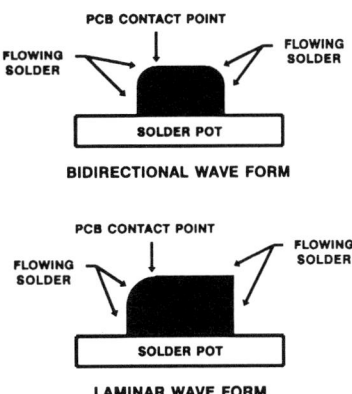

FIGURE 9.10. Bi-Directional Wave Form versus Laminar Wave Form.

decreases the surface tension of the molten solder, which produces a smaller contact angle with the PCA. The smaller contact angle decreases the size of the web that forms between the solder and the PCA. This effect also allows the solder to pull itself away from the PCA more deliberately.

Another variable which helps to decrease the size of the web formed between the PCA and the solder is the entry angle, which is the angle of contact between the PCA and the solder wave. This angle is achieved by adjusting the PCA transport system. Entry angles are typically set at 4° to 7°.

The wave penetration depth is important if the plated through-holes are to be properly filled with solder. The variable here is PCB thickness. As the PCB becomes thicker it becomes more difficult to fill the plated through-holes. As a general rule, PCBs up to approximately 1.5mm (0.062″) thick should penetrate into the wave half of their thickness, while PCBs up to approximately 2.5mm (0.100″) thick should penetrate into the wave three-quarters of their thickness.

Wave soldering oils have been used to decrease surface tension and dross formation. Oil can be injected into the solder wave, resulting in a solder–oil intermix, to reduce surface tension, or it can be used to reduce dross formation by covering exposed solder. These techniques were developed to improve the bidirectional waveform. However, with the development of the unidirectional waveform and controlled entry angles, the solder–oil intermix is unnecessary. Oil can still be used to cover exposed solder, but it may not be worth the trouble.

A hot air knife is available as an option from one supplier. This device is located directly after the solder module. It directs a stream of high-velocity hot air at the solder joints on the bottom of the PCA. Its purpose is to remove solder bridges and weak solder joints that have not wet properly, using the force provided by the directed hot air.

9.5.2 Dual Wave Approach

The dual wave system was developed to wave solder surface mount components. There are two waveforms, a turbulent wave and a laminar wave, as displayed by Figure 9.11. The turbulent wave is used to remove any trapped flux gas bubbles and form the initial solder joint. A turbulent wave tends to produce an oversized, irregular-shaped solder joint. The final, properly shaped solder joint is generated by the laminar wave.

A turbulent wave is created by pumping solder through a nozzle that incorporates either a narrow opening or a series of small holes. This creates a high-pressure, turbulent waveform that can remove trapped flux gas bubbles and enter dense areas. It also allows the solder to contact the land, termination, or lead at good wetting angles, something that is difficult to do

FIGURE 9.11. Example of a Laminar Wave with a Turbulent Wave.

with a laminar wave, as noted earlier. Most wave soldering systems direct the turbulent wave toward the laminar wave, which promotes a good scrubbing action. One design also moves the turbulent wave in a back-and-forth, sideways motion in an attempt to improve the scrubbing action.

Some dual wave systems use individual solder pots for each wave, while other systems draw solder from a single pot. Each wave has its own pumping capability as well. This allows the laminar wave to be operated without the turbulent wave. Turbulent waves expose considerably more solder to the atmosphere than a laminar wave. This produces significantly more dross. As a result, turbulent waves may require more maintenance and cleaning than just a laminar wave.

9.5.3 Vibrating Wave Approach

The vibrating wave, created by one supplier, was also developed to solder surface mount components. It consists of a single laminar wave that is agitated by ultrasonically vibrating the wave. This agitation generates small solder bubbles that can remove trapped flux gas bubbles and enter dense areas. It also allows the solder to contact the land, termination, or lead at good wetting angles. The single laminar wave is divided into two zones. The first zone is agitated; the second zone is stationary. The first zone removes any trapped flux gas bubbles and forms the initial solder joint; the second zone produces the final, properly shaped solder joint.

In terms of the wave soldering system itself, the vibrating wave concept does have advantages over the dual wave concept. A smaller solder pot can be used, and dual heating and pumping systems are not required. A single vibrating wave also produces less dross than a dual wave.

9.5.4 Cool Down

Cool down affects the final solder joint integrity and strength. Solder that is cooled at a reasonable rate achieves a small, fine-grain structure. This type of grain structure provides a strong, reliable solder joint. Cool-down rates of 1 to 2°C/second (1.8° to 3.6°F/second) are generally preferred. Small- and medium-sized PCAs may cool down properly without forced cooling. Large, multilayer PCAs may require forced cooling provided by fans located after the solder module.

9.5.5 Dross Control

Dross control is important from a safety and cost standpoint. Dross, because it contains lead, is a safety concern. Operators should always wear protective masks or respirators and gloves when handling dross. Dross containers should be kept tightly sealed at all times. Operators should be instructed to wash their hands before drinking, eating, or smoking.

Dross generation can be reduced by proper wave geometry design. Solder alloy and temperature also play a role in dross generation. As noted earlier, a coating of oil can be used to cover the solder pot. This, however, does little to reduce gross generation, as most of the dross is generated by the solder wave, especially the more turbulent waves. The lead-saturated oil is a waste disposal concern. The oil fumes also deposit an oily film on the equipment and exhaust system. The use of oil must be balanced against disposal and maintenance issues. Chemicals can be used to release molten solder that is suspended in the dross. This allows the trapped solder to return to the solder pot.

9.5.6 Inert Atmosphere Soldering

Wave soldering equipment that uses an inert atmosphere is now available. An inert atmosphere system, using nitrogen, provides an oxygen-free soldering environment. The flux, preheat, and solder modules are enclosed in an airtight chamber. Thus, the PCA is only subjected to an inert environment after fluxing. This allows the use of a less active (low solids or no clean) flux. An inert atmosphere also reduces the surface tension of the molten solder and decreases dross generation.

9.6 PCB HANDLING

The transport system is important not only because it carries the PCA through the wave soldering system, it also presents the PCA to the various modules (flux, preheat, solder) in the proper position and orientation. Two methods are used to transport a PCA through the wave soldering system: a finger conveyor or a pallet conveyor.

The finger conveyor is the most common method in use today because of its low cost and flexibility. However, it does not support the PCB as well as a pallet conveyor. It consists of two parallel drive chains to which the fingers are attached. The fingers grip the PCA on two opposite edges. Many different finger designs are in use. The V groove and L shape, as shown in Figure 9.12, are the most common. In general, they are made from titanium. They are made to contact as little of the PCB edge as possible. It is a good

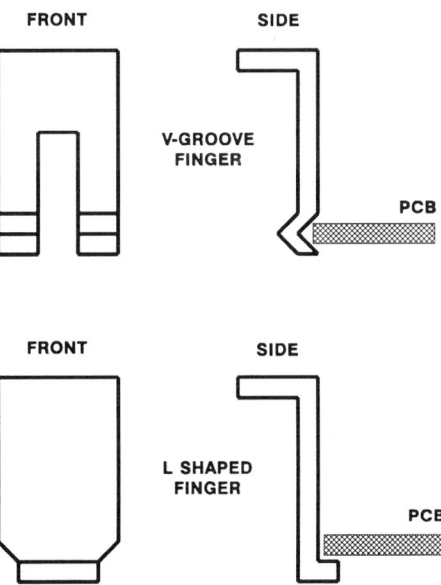

FIGURE 9.12. Common Wave Soldering Fingers.

182 Applied Surface Mount Assembly

practice to have a minimum 3.18mm (0.125") border along two parallel edges (preferably the two longest edges) that are free of components and circuitry. The fingers are fabricated with a slight outward taper. This allows them to apply moderate pressure to the PCB edge, which keeps the PCB firmly seated in the finger. When adjusting the width of the finger conveyor it is very important to achieve the correct width. If the finger conveyor is adjusted too wide the PCB may fall off during transport. If the finger conveyor is adjusted too tight the PCB may bow upward or downward. It is critical that the PCB be kept flat during wave contact. Remember that when the PCB is heated above its glass transition temperature (Tg) it will expand. The finger conveyor must be designed and adjusted properly to compensate for this expansion. An automatic finger cleaner, which removes flux build-up, is a valuable option.

The pallet conveyor provides better PCB support because a fixture is used to hold the PCB while it is transported through the wave soldering system. Once again, two parallel drive chains are used, this time to transport the pallet. Pallets can be fabricated from phenolic laminates, anodized aluminum (do not use bare aluminum) and titanium. Either an adjustable or dedicated fixture can be used. A combination locating and drive pin is usually positioned in each corner of the fixture. These pins are inserted into the drive chain. Pallet conveyors are effective, but more expensive than finger conveyors, because pallets must be designed and fabricated. The pallets also require frequent cleaning to remove flux. Also, once the wave soldering process is complete, the pallet must be moved back to the beginning of the process.

A complete wave soldering system is illustrated in Figure 9.13 and pictured in Figure 9.14.

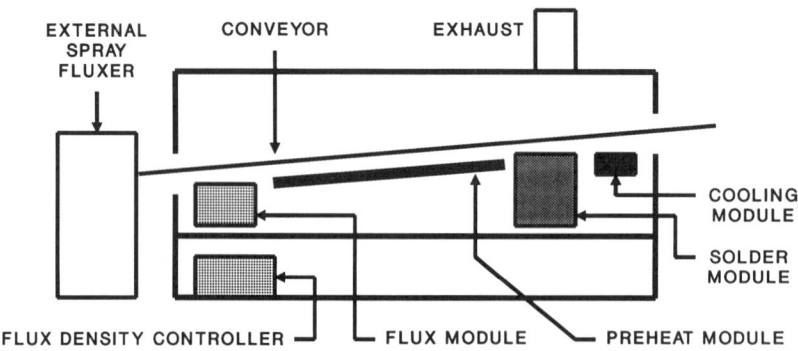

FIGURE 9.13. Wave Soldering System Concept.

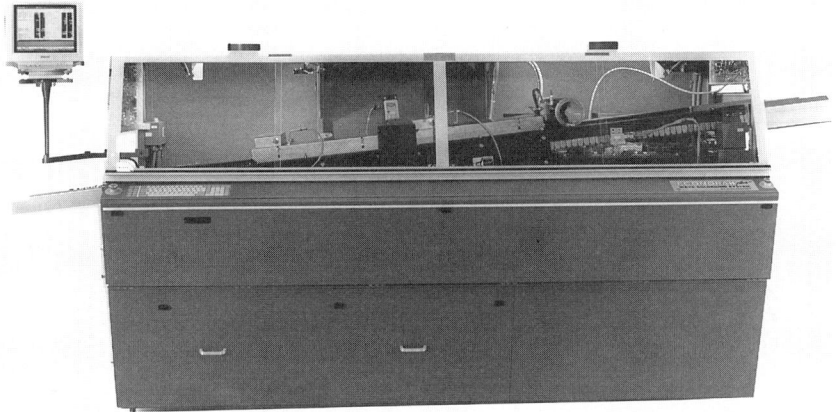

FIGURE 9.14. Example of a Wave Soldering System.

REFERENCES
1. Bernier, Dennis F. "Understanding the Controls Necessary for Wave Soldering Reliability." *Electronic Manufacturing*, August 1988, pp. 35–36.
2. Bilotta, Anthony J. *Connections in Electronic Assemblies*, New York: Marcel Dekker, 1985.
3. Botham, Robert M., Lowell, Charles R., and Sterritt, Janet R. "Wave Soldering Mixed Technology Boards." *Electronic Packaging and Production*, November Supplement, 1990, pp. 28–30.
4. Hymes, Les. *Cleaning Printed Wiring Assemblies in Today's Environment.* New York: Van Nostrand Reinhold, 1991.
5. Manko, Howard H. *Solder Handbook for Printed Circuits and Surface Mounting*, New York: Van Nostrand Reinhold, 1986.
6. Prasad, Ray P. *Surface Mount Technology—Principles and Practice*, New York: Van Nostrand Reinhold, 1989.

10
Cleaning

GLOSSARY

Aqueous Cleaning   A PCA cleaning process that uses water to remove water soluble flux or water and a saponifier to remove rosin flux.

Freeboard   A ratio applied to solvent cleaners. A certain amount of freeboard is required to reduce solvent loss. It is determined by dividing the vertical distance from the top of the vapor zone to the top of the cleaner by the narrowest dimension of the opening. Cleaners using CFCs typically have a freeboard of 75% (0.75:1), while HCFC cleaners require a freeboard of 100% (1:1).

Ionic Contamination   Contamination that contains molecules that may form positively and negatively charged particles (ions) when in solution.

Ozone Depletion Potential (ODP)   The ozone depletion potential of a chemical. The standard base line chemical is CFC-11, which has an ODP of 1.0.

Polar Contamination   Contamination that contains molecules that have a dipole moment, which means the positively charged particles do not have the same center (pole) as the negatively charged particles. When exposed to certain solvents some polar molecules dissociate to form ions.

Saponifier   Alkaline chemical agents that convert most of the elements in a rosin flux into a soap. The soap is then washed away with water. The saponifier is usually added to the first one or two cleaner modules.

Semiaqueous Cleaning   A PCA cleaning process that uses a hydrocarbon-based solvent to remove rosin flux. The solvent is then removed with a water wash.

Solvent Cleaning   A PCA cleaning process that uses a chlorinated or fluorinated solvent to remove rosin flux.

Surfactants   Chemical additives that lower the surface tension of water. Stands for surface-active agents.

10.0 INTRODUCTION

Cleaning is truly a process in turmoil. The surface mount industry has relied almost entirely on chlorofluorocarbon (CFC) solvents to clean surface mount assemblies. However, in recent years it has been determined that CFC-type solvents damage the earth's ozone layer. The Montreal Protocol, a document mandating the phase-out of CFC and other ozone-depleting chemicals, has been ratified by most nations. It has established specific guidelines for the phase-out and eventual elimination of most ozone-depleting chemicals. The initial and current production phase out plan for CFC materials is shown below.

October 1987 Base Line	June 1990 Base Line
1989: Reduce to 1986 level	1993: Reduce by 20%
1994: Reduce by 20%	1995: Reduce by 50%
1998: Reduce by 50%	1997: Reduce by 85%
	2000: Eliminate use

The elimination of CFC-based cleaning has created a mad dash to find other materials and processes capable of removing contamination. As it stands today, the available cleaning processes can be divided into three categories: solvent, aqueous, and semiaqueous. Operator safety and environmental concerns should always be addressed when using solvents. Information can be obtained from local, state, and federal government agencies, as well as material and equipment suppliers.

The compatibility of components with the cleaning process is always a concern. Unsealed components can be damaged or destroyed by solvent penetration. Certain plastics can also be damaged by some of the solvents. Component and solvent compatibility should always be verified.

This chapter will start by reviewing the contaminants that need to be removed, and then each of the three cleaning processes will be discussed, including solvents and equipment. Methods for testing the cleanliness of the PCA will also be reviewed. Table 10.1 provides a summary of the various cleaning methods.

10.1 CONTAMINANTS

Before establishing a cleaning process it must be clearly understood what needs to be removed from the PCA. During fabrication and assembly the PCA is exposed to a wide spectrum of contaminants. These contaminants can

TABLE 19.1 Cleaning Process Summary

Description	Solvent (CFC)	Aqueous	Aqueous with Saponification	Semiaqueous
Equipment cost:				
Batch	Moderate	Low	Low	Moderate
Inline	Moderate	Low to moderate	Low to moderate	High
Equipment size:				
Batch	Medium	Small	Small	Medium
Inline	Medium	Large	Large	Large
Material cost	High	Low	Moderate	High
Process cost	High	Low	Moderate	High
Flux type	Rosin	Water soluable	Rosin	Rosin
Flux residue	Inactive (RMA)	Active (WSF)	Inactive (RMA)	Inactive (RMA)
Cleaning ability	Very Good	Excellent	Good	Excellent
PCA drying	Easy	Difficult	Difficult	Difficult
Toxicity	Low	Low	Low	Low
Attributes	Surface tension One step process Compatibility	Low cost Availability One step process	Availability	Surface tension Availability
Concerns	Phase out by 1995 High cost Availability Environment	Surface tension Waste disposal Water usage Compatibility	Surface tension Two step process Waste disposal Water usage Compatibility Caustic residue	Two step process Waste disposal Water usage Compatibility Combustible

be placed in four groups: ionic, nonionic, polar, and nonpolar. For best results, the PCA should always be cleaned within one hour of soldering.

Ionic molecules form positively and negatively charged particles (ions) when in solution. Polar molecules have a dipole moment, which means the positively charged particles do not have the same center (pole) as the negatively charged particles. Some polar molecules, when exposed to certain solvents, dissociate to form ions. Ionic and polar contamination are a concern because their presence can cause electrical leakage between conductors, dielectric breakdown, serious corrosion, and dendritic growth—all of which affect the reliability of an electrical assembly.

Nonionic molecules do not form ions under any conditions. A nonpolar molecule has no dipole moment. Nonionic and nonpolar contamination are not as serious as ionic and polar contamination, but they can have a negative effect on an electrical assembly. These contaminants can prevent good electrical contact between the probes on a bed-of-nails test fixture and the electrical assembly. It can also cause poor electrical contact between connectors. Contamination can also affect how well a conformal coating adheres to the surface of an electrical assembly. And finally, the contamination may attract other contamination to the electrical assembly, such as dust and dirt or, in the worst case, ionic and polar contamination.

10.2 SOLVENT CLEANING

Solvent cleaning, as noted above, is on the decline because of concerns about the solvent's impact on the environment. New solvents have been proposed to replace the solvents currently in use; however, it appears that their use may also be curtailed because of the same environmental concerns. Until recently, solvent cleaning has been the method of choice to remove the rosin-based fluxes commonly used in solder paste.

The basic fundamentals of solvent cleaning are the same whether a batch or in-line system is used. When a PCA enters the system the hot vapor condenses on the PCA, beginning the cleaning process. If additional cleaning is required the PCA can be sprayed with solvent and/or completely immersed in the solvent. Internal or external stills can be used to remove contamination from the solvent; however, external stills provide better solvent control, and they can handle larger volumes of contamination. Solvent conservation methods, such as designing the PCA to prevent solvent dragout and using solvent recovery systems, should be actively employed to reduce solvent loss.

10.2.1 Solvents

The two most common solvents used in the United States for the removal of rosin-based fluxes have been 1, 1, 1, trichloroethane (methyl chloroform), a nonpolar chlorinated solvent, and CFC 113, a nonpolar fluorinated solvent. Chlorinated solvents have a higher boiling point than fluorinated solvents, approximately 75°C (167°F) versus 40°C (104°F). A note of caution, chlorinated solvents are generally not compatible with PVC piping.

Most contaminants are a combination of ionic, nonionic, polar, and nonpolar materials. The polar and nonpolar contaminants are removed most effectively by mixtures of polar and nonpolar solvents, called azeotropes. An azeotrope is simply a blend of two or more solvents that behave as a single compound, even when boiled and condensed. The boiling point of the blend is lower than any of the solvents that were added to the mixture. Azeotrope solvents can be chlorinated or fluorinated. The military has promoted the use of an aqueous rinse after solvent cleaning to remove ionic and nonionic contamination. However, in most cases, solvent cleaning alone will produce satisfactory results.

Another class of solvent, hydrochlorofluorocarbons (HCFC), is being proposed as a replacement for the CFC solvent. However, at the time of this writing it appears that HCFCs may not be introduced at all or their introduction and life span will be very limited. The HCFC solvents have a very low boiling point, approximately 30°C (86°F). Special handling and equipment is required to prevent solvent loss. Systems designed to use CFC solvents cannot be used with HCFC solvents. Because of the low boiling point the HCFC solvents require additional freeboard.

10.2.2 Batch Solvent Cleaners

Batch solvent cleaners are used mainly for low-volume cleaning applications. If still available, batch cleaners can be purchased for $10,000 to $30,000. The main elements of most batch cleaners, as shown in Figure 10.1, are the main tank, with one or two sumps, condensing (cooling) coils, a spray wand, and a manual or automated hoist. A system with two sumps, one boil sump and one rinse sump, is preferred. Clean solvent from the condensing coils is returned to the rinse sump, which overflows into the boil sump. The boil sump generates the solvent vapor. A cover should be placed over the main tank when the cleaner is not in use to minimize solvent loss.

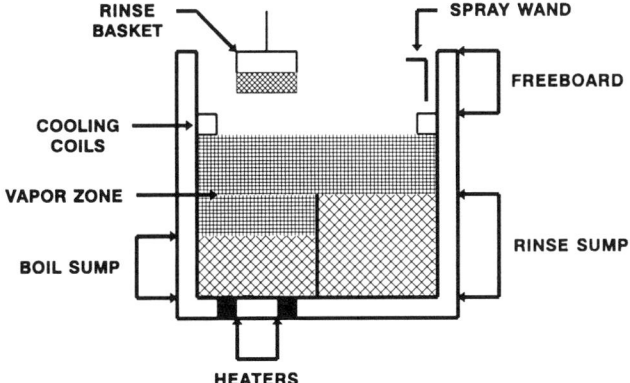

FIGURE 10.1. Batch Solvent Cleaning System Concept.

The PCA is lowered into the vapor blanket to begin the cleaning cycle. The vapor condenses on the cooler PCA, which begins the process of contamination removal. The PCA may also be sprayed with the wand or immersed in the rinse sump.

Batch cleaners are very dependent on the operator, especially if a manual hoist is used. To minimize solvent loss an automated hoist is highly recommended. For best results, the PCA should be cleaned within one hour of soldering. The PCA should be placed in the vapor for at least two minutes. If additional cleaning is required, spray the PCA with the wand for approximately one minute (the spray nozzle should be well below the vapor line) or immerse the PCA in the rinse sump for approximately one minute. Remove the PCA from the vapor slowly to minimize solvent loss.

10.2.3 In-line Solvent Cleaners

In-line solvent cleaners are intended for medium- to high-volume applications, or situations where better control of the cleaning process is required. These cleaners, if still available, cost from $75,000 to $125,000. They operate much the same as batch solvent cleaners. However, the cleaning process is much more automated. It is a more stable process than batch cleaning because the operator-induced variables have been eliminated. A typical in-line solvent cleaner, as shown in Figure 10.2, consists of five different zones: entrance vapor zone, preclean spray zone, spray/immersion zone, distillate spray zone, and exit vapor zone. There are numerous variations on this basic configuration. Four sumps are usually used: a boil sump, a preclean sump, a spray/immersion sump, and a distillate sump. The solvent cascades from the distillate sump to the boil sump. Condensing (cooling) coils are used to contain the vapor. The PCA is transported through the cleaner on a stainless steel mesh belt, which is inclined on both ends and horizontal in the center. Plastic or metal cleats are commonly attached to the mesh belt to prevent the PCA from sliding down the incline. The spray nozzle angle, pattern, and pressure can be varied in most systems. Spray pressure varies from 25 to 200 PSI, depending on the system. The top-side spray pressure should always be set slightly higher than the bottom-side spray pressure. This will force the PCA against the mesh belt and prevent the PCA from being tossed around.

The cleaning process is started by placing the PCA on the mesh belt, which slowly transfers it into the entrance vapor zone where the vapor condenses on the PCA. The PCA is then sprayed top and bottom in the preclean spray zone. The PCA is completely covered with solvent in the spray/immersion zone. The overhead spray is used to agitate the solvent. As a final rinse, the

Cleaning 191

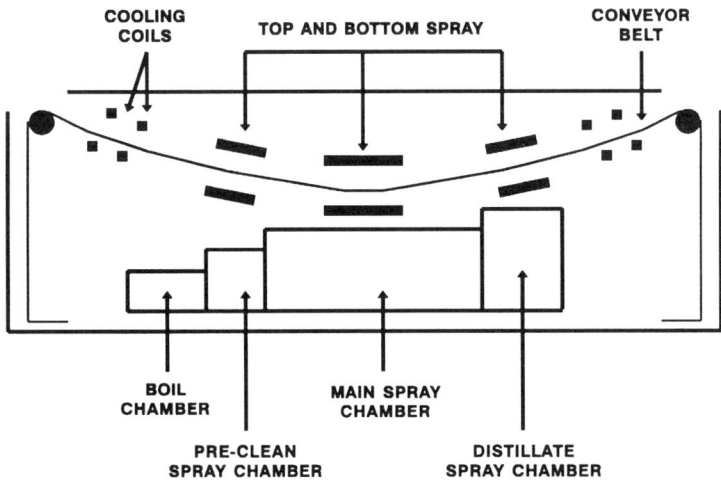

FIGURE 10.2. Continuous Solvent Cleaning System Concept.

distillate spray zone sprays the top and bottom of the PCA with very clean solvent. The PCA then passes through the exit vapor zone and out of the cleaner. The transport speed ranges from 120 to 300 cm/minute (4 to 10'/minute), depending on the size of the cleaner, the size and density of the PCA, and the amount of contamination to be removed.

10.3 AQUEOUS CLEANING

Aqueous cleaning has been popular for removing water soluble fluxes from through-hole PCAs. Aqueous/saponifier cleaning has also been used, to a lesser extent, to remove rosin flux from through-hole PCAs. It has only recently been considered for removing contamination from surface mount assemblies, preferably without saponification. The first step toward making aqueous cleaning a viable process for surface mount was the development and introduction of water soluble solder pastes. Proven, reliable water soluble solder pastes are available today from most solder paste suppliers. The residue from a water soluble solder paste is easier to remove than a rosin-type residue; however, the concern with aqueous cleaning is the ability to penetrate under components that have minimal clearance between the bottom surface of the component and the top surface of the PCB. The residue from a water soluble solder paste must be removed from the PCA. Aqueous cleaner design has advanced considerably in the past several years.

Water includes certain compounds that make it hard. Water softeners can be added to improve cleaning efficiency and prevent the build-up of these compounds in the cleaner. Deionized water may also be used when very pure water is required. Surfactants can be added to reduce the surface tension of the water. Hot water, typically 50°C to 65°C (125°F to 150°F), promotes cleaning.

10.3.1 Batch Aqueous Cleaners

Batch aqueous cleaners are also used principally for low-volume cleaning applications. A batch aqueous cleaner, as shown in Figure 10.3, can be purchased for approximately $5,000 to $20,000. It contains a main tank, a basket for holding the PCAs during cleaning, an overhead spray system, and a convection dryer. This system uses a programmable controller to establish prerinse, wash, and postrinse times, number of wash cycles, and drying time.

FIGURE 10.3. Example of a Batch Aqueous Cleaning System.

If a saponifier is used it will add and monitor the chemicals. One advantage of this type of system is that the PCAs are cleaned in a vertical position, which allows the water and contaminants to drain off of the PCA more efficiently. Since the volume of water on the surface of the PCA is kept to a minimum, drying is also easier. Puddling of the water on the surface of the PCA is one of the problems associated with aqueous cleaning.

The cleaning cycle is quite simple. The PCAs are placed in the basket and the door is closed. The controller is programmed with the proper information. A quick prerinse is used to remove most of the contamination. This is followed by one or more extended wash cycles and a postrinse. Lastly, the convection dryer circulates hot, dry air through the cleaner to dry the PCAs.

10.3.2 In-line Aqueous Cleaners

In-line aqueous cleaners, priced from around $50,000 to $100,000, are intended for medium- to high-volume applications. Unlike solvent cleaners, an aqueous batch cleaner may actually provide better process control than an in-line aqueous cleaner. In-line cleaners consist of various modules that clean and dry the PCA as it progresses through the cleaner. A typical in-line aqueous cleaner, as shown in Figure 10.4, consists of at least five different modules: prerinse with air knife, wash #1 with air knife, wash #2 with air knife, postrinse with air knife, and final dry. Since vapor loss is not a concern, no freeboard is required, which means a horizontal stainless steel mesh belt can be used. The spray nozzle angle, pattern, and pressure can be varied in most systems. Spray pressures of approximately 30 PSI are typically used, but pressures of 100 PSI have been tried. The top-side spray pressure should always be set slightly higher than the bottom-side spray pressure. This will

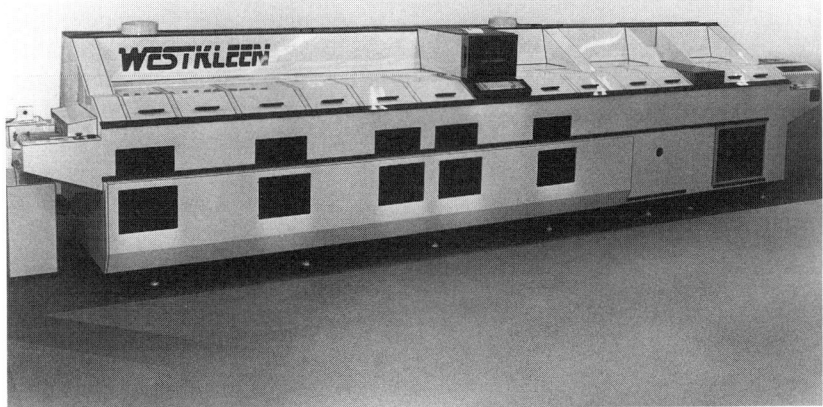

FIGURE 10.4. Example of a Inline Aqueous Cleaning System.

force the PCA against the mesh belt and prevent the PCA from being tossed around.

The cleaning process is started by placing the PCA on the mesh belt, which slowly transfers it into the cleaner. The PCA is sprayed top and bottom in each module. Excess water is removed between modules with an air knife. Final drying occurs in the dryer module. Since the PCA is in a horizontal position, the water tends to puddle on the surface, this makes drying much more difficult. One design holds the PCA in a vertical position during cleaning, which makes cleaning and drying easier, but PCA handling is much more difficult. Transport speeds range from 120 to 300 cm/minute (4 to 10'/minute), depending on the size of the cleaner, the size and density of the PCA, and the amount of contamination to be removed.

10.4 SEMIAQUEOUS CLEANING

10.4.1 Solvents

Numerous semiaqueous solvents, intended for the removal of rosin-based fluxes, are available today. Each one is a specific formulation. Detailed information about these proprietary solvents should be obtained from the solvent supplier. Most semiaqueous solvents are hydrocarbon-based formulations. Suppliers claim that these solvents are generally biodegradable, nontoxic, and noncorrosive. Since these solvents contain no chlorine their ozone depletion potential (ODP) is zero. However, most of them are classified as a volatile organic compound (VOC).

Semiaqueous solvents have been identified as generally having a superior cleaning capability when compared to CFC 113. A concern related to these solvents has been their low flashpoint. New formulations are being developed to address this issue, along with improvements in cleaning equipment. Some plastics and elastomers used on PCAs may be affected by these solvents. After application of these solvents, the PCA should be thoroughly washed with water to remove the solvent. Surfactants can be added to the water to reduce the surface tension.

10.4.2 Batch Semiaqueous Cleaners

Batch semiaqueous cleaners are used primarily for low-volume cleaning applications. A batch semiaqueous cleaning system, available for about $25,000, is shown in Figure 10.5. Two units are used with this set-up, one for the solvent wash and the other for the aqueous rinse. Both cleaners contain a main tank, a basket for holding the PCAs during cleaning, and an overhead

FIGURE 10.5.   Example of a Batch Semiaqueous Cleaning System.

spray system. The aqueous cleaner also contains a convection dryer. Semiaqueous solvents are flammable, so some method of fire control, such as an inert atmosphere, is required. Both systems use a programmable controller to establish prerinse, wash, and postrinse times, number of wash cycles, and drying time. Both of these cleaners hold the PCAs in a vertical position, an advantage because the solvent, water, and contaminants drain off of the PCA more efficiently. The volume of water on the surface of the PCA is kept to a minimum, which makes drying easier. Puddling of the water on the surface of the PCA is a problem affiliated with aqueous cleaning.

196   Applied Surface Mount Assembly

The cleaning cycle begins by placing the PCAs in the semiaqueous cleaner. The controller is programmed with the proper information. A short prerinse is followed by one or more extended wash cycles and a postrinse. The PCAs are then placed in the aqueous cleaner. Again, the controller is programmed with the proper information. A series of prerinse, wash and postrinse cycles are performed to remove the solvent. The PCAs are dried by circulating hot, dry air through the cleaner.

10.4.3 In-line Semiaqueous Cleaners

In-line semiaqueous cleaners, which can be quite expensive at around $125,000 to $200,000, are intended for medium- to high-volume applications. As with aqueous cleaners, the batch semiaqueous cleaner may offer better process control than an in-line semiaqueous cleaner. In-line semiaqueous cleaners, as pictured in Figure 10.6, are actually two cleaners in one. The first half of the cleaner is devoted to washing the PCA with the semiaqueous solvent, the second half of the cleaner uses water to remove the solvent. The solvent half consists of one or more modules that apply solvent to the PCA. A major concern here is the flashpoint of the solvent. The solvent mist is of particular concern. One method of fire control is to use an inert atmosphere inside the solvent module; another method is to use a spray-under-immersion concept to prevent the mist from occurring. An

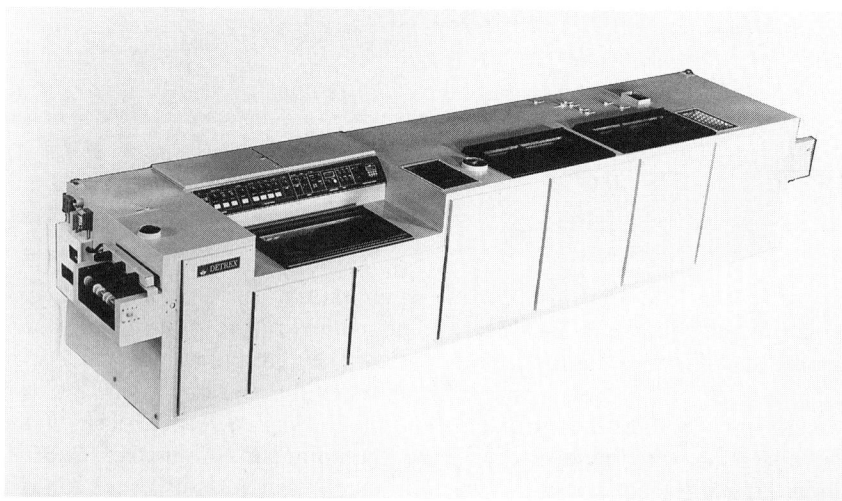

FIGURE 10.6.   Example of an Inline Semiaqueous Cleaning System.

air knife is used to remove as much solvent as possible before passing the PCA into the aqueous wash. As with in-line aqueous cleaners, a series of modules consisting of spray nozzles and air knives is used to wash the PCA with water. Vapor loss is not a concern, so no freeboard is required, which allows a horizontal stainless steel mesh belt to be used. The spray nozzle angle, pattern, and pressure can be varied in most systems. Typical spray pressures are approximately 30 PSI. The top-side spray pressure should always be set slightly higher than the bottom-side spray pressure. This will force the PCA against the mesh belt and prevent the PCA from being tossed around.

The PCA is placed on the mesh belt and transferred into the solvent module, where the PCA is sprayed top and bottom with solvent. An air knife is used to remove excess solvent before transferring the PCA into the aqueous module. One or more aqueous modules spray the top and bottom of the PCA with water. Excess water is removed with an air knife. Final drying occurs in the dryer module. With the PCA in a horizontal position, the water tends to puddle on the surface, which makes drying more difficult. Transport speeds can vary from 120 to 300 cm/minute (4 to 10'/minute), depending on the size of the cleaner, the size and density of the PCA, and the amount of contamination to be removed.

10.5 WASTE WATER MANAGEMENT

With the increased use of water in the cleaning process, water pretreatment (preparing industrial waste water for discharge into a public treatment system) is becoming a growing concern. Pretreatment concerns include water temperature, pH level, heavy metal content, chemical oxygen demand (COD), biological oxygen demand (BOD), total toxic organics (TTO), and other contaminants such as oil and grease. Specific requirements vary from location to location. Pretreatment equipment may have to be purchased, in addition to the cleaning equipment, to provide proper waste water management. Closed loop aqueous recycling systems are available to minimize the discharge of water.

10.6 CLEANLINESS TESTING

A constantly asked question is "How clean is clean?" Many attempts have been made to answer this question, especially by the military. The answer depends, to a great extent, on the application and environment for which the PCA is designed. A PCA intended for use in the aerospace industry will have more stringent cleanliness requirements than a PCA intended for use in a television set. Four common methods used for evaluating cleanliness are discussed here.

10.6.1 Visual Inspection

Visual inspection is done by a skilled operator using $2\times$ to $25\times$ magnification. Contamination on the surface of the PCA can usually be seen; however, it is impossible to see contamination under components unless they are removed for inspection. The results of visual inspection are strictly subjective.

10.6.2 Solvent Extraction Testing

Solvent extraction testing comprises immersing the PCA in a test solution of 75% isopropyl alcohol and 25% deionized water, which removes the contaminants and then measures their ionic conductivity. The result is displayed in micrograms of sodium chloride (NaCl) per square unit of PCB area. This method basically averages the amount of contamination over the PCB surface. MIL-P-28809, "Printed Wiring Assemblies," defines the expected result based upon the test equipment used. This method is only accurate if all of the contamination is removed from under each component. There is some question as to how successfully this can be done on a surface mount assembly, since the components are much closer to the surface of the PCB. New equipment is available that heats and agitates the solution. This does a much better job of removing contaminants from under surface mount components. Also reference IPC-TM-650 method 2.3.26.1.

10.6.3 Surface Insulation Resistance (SIR) Testing

Surface insulation resistance testing determines the presence of contamination in a specific location on the PCB. Basically, this method determines the resistance of the surface of a dielectric material between parallel conductors in a PCB test pattern. The test measures the current leakage caused by surface contamination. The resistance should be quite high unless contamination on the surface of the PCB causes it to drop. Test procedures can be found in several documents, including IPC-SF-818, "General Requirements for Electronic Soldering Fluxes," and Bellcore TA-NWT-000078, "Generic Physical Design Requirements for Telecommunication Products and Equipment." The desired SIR-results depend on the test conditions. Testing is done under specific temperature and relative humidity conditions.

10.6.4 Electromigration Testing

Electromigration testing determines the probability of surface contamination causing dendritic growth and current leakage between two parallel conductors on the surface of a PCB. Dendritic growth can occur between two parallel conductors when the proper bias voltage and relative humidity is applied in the presence of surface contamination. Test methods are described in Bellcore TA-NWT-000078, "Generic Physical Design Requirements for Telecommunication Products and Equipment."

REFERENCES
1. Andrus, James J. "Managing PWA Aqueous Cleaning Pretreatment." *Circuits Assembly*, December 1990, pp. 59–61.
2. Attalla, Gary. "Semiaqueous: A Progress Report." *Circuits Manufacturing*, April 1990, pp. 22–26.
3. Banks, Sherman. "Eliminating CFCs," *Printed Circuit Assembly*, August 1990, pp. 31–35.
4. Brinton, James B. "CFC Alternatives Reach Critical Mass," *Circuits Manufacturing*, April 1990, pp. 14–16.
5. Elliott, Donald A. "Cleaning Surface Mount Assemblies." *Electronic Packaging and Production*, June 1990, pp. 46–49.
6. Hymes, Les, *Cleaning Printed Wiring Assemblies in Today's Environment*. New York: Van Nostrand Reinhold, 1991.
7. IPC-SF-818. "General Requirements for Electronic Soldering Fluxes." IPC, Lincolnwood, IL, March 1988.
8. Merrithew, Larry. "CFC Alternatives: Where Do We Go From Here?" *Surface Mount Technology*, December 1990, pp. 22–23.
9. MIL-P-28809. "Printed Wiring Assemblies." DOD, Washington, D.C., October 1981.
10. Prasad, Ray P. *Surface Mount Technology—Principles and Practice*, New York: Van Nostrand Reinhold, 1989.
11. Samsami, Darius. "Cleaning Without CFCs: The Semiaqueous Solutions." *Electronic Packaging and Production*, June 1991, pp. 62–68.
12. Sullivan, James. "PCB Cleaning: Making the Intelligent Choice." *Printed Circuit Assembly*, August 1990, pp. 11–17.
13. TA-NWT-00078. "Generic Physical Design Requirements for Telecommunications Products and Equipment." Bellcore, Morristown, NJ, December 1990.

11
Rework

GLOSSARY

Contact Soldering Heating of the solder joint is accomplished by direct contact with the solder joint. A heated tip or collar is used to contact the solder joint and/or leads.

Defect A solder joint that deviates from the established workmanship standard.

Hot Gas Soldering Heating of the solder joint is accomplished by hot air. A nozzle is used to direct the flow of heated air.

Infrared Soldering Heating of the solder joint is accomplished by focused infrared energy. This type of heating is intended primarily for large integrated circuit packages.

Thermal Shock A condition which occurs when the temperature of the material is increased too quickly.

11.0 INTRODUCTION

Unfortunately, in the real world defective solder joints occur. These defects must be reworked to bring the solder joint to an acceptable condition. Rework must be done correctly when it is required. However, rework should be viewed as an evil process because it does not add value to the PCA. Rework is a process that requires excellent operator skills. An unskilled operator can quickly create a reliability nightmare. When equipped with the proper tools and training, an operator should find it easier to rework surface mount components than through-hole components. Removing a lead from a plated through-hole is more difficult and stressful to the printed circuit board (PCB) than removing a termination or lead from a land. Surface mount rework is sometimes more challenging because of the smaller lead pitches and higher lead counts. Plated through-holes and lands are easily damaged during the rework process. Care must be taken not to overheat them. Lifted pads and lands are a common result of overheating.

One of the biggest transgressions in manufacturing is unnecessary rework. Acceptable solder joints are often reworked because people cannot tell the difference between an acceptable solder joint and an unacceptable solder joint. A well written workmanship standard and good training are required if unnecessary rework is to be avoided.

Rework is done using one of three methods: contact soldering, hot gas soldering, and infrared soldering. There are attributes that are common to all rework equipment, but the most important is the ability to heat the solder joint or joints to the proper liquidus. This chapter will shed some light on the equipment and methods that are available, both manual and automated, and the concerns and attributes of each.

11.1 COMPONENT CONCERNS

11.1.1 Thermal Shock

Some surface mount components are more delicate than others. One of the main problems that occurs during rework is damage from thermal shock, especially with multilayer ceramic capacitors. Care must be taken to heat capacitors slowly and uniformly. The temperature ramp rate should be kept around 5°C/second (9°F/second). Today's capacitors are more robust than earlier capacitors. Suppliers have continually upgraded their manufacturing capability, resulting in better components. Better rework equipment has also helped decrease thermal shock.

11.1.2 Component Terminations and Leads

Component lead and termination finish is an important factor that will affect solderability and reliability. See Chapter 5 for additional information.

Component leads are typically made from copper and copper alloys. Other alloys, such as Kovar (53% iron, 29% nickel, 17% cobalt) and Alloy 42 (42% nickel, 58% iron), are also used, but to a lesser extent. The main concern during the rework process is poor wetting caused by copper/tin intermetallics. If the lead is heated above liquidus for too long the intermetallic layer will grow to the extent that the operator is trying to solder to the copper/tin intermetallic, rather than the tin/lead solder. Copper/tin intermetallic will not wet properly, causing a dewetted solder joint.

Ceramic rectangular and cylindrical components may suffer from leaching of the termination while above liguidus. During the soldering process the adhesion layer metallization is vulnerable to dissolution in tin. This problem can be prevented by adding a barrier layer of nickel between the adhesion layer and the outer coating. Another, less effective, method can also be used to prevent leaching. In this method a solder alloy containing 2% silver (62Sn/36Pb/2Ag) is used instead of the 63Sn/37Pb alloy.

11.2 CONTACT SOLDERING

Contact soldering is accomplished by making direct contact with the solder joint, using a heated metal tip or collar that is attached to a soldering iron. Soldering tips are used to heat individual solder joints and/or leads, while soldering collars are used to heat multiple solder joints and/or leads at one time.

There are numerous designs for single tip soldering irons. Various removable tips are available for particular heating requirements. Standard soldering irons should be preset to operate in a temperature range of 290°C to 370°C (554°F to 698°F). See Figure 11.1 for the relationship between the soldering iron tip temperature and the plated through-hole temperature. Avoid applying heat to a pad or land for more than three seconds.

There are also numerous designs for collar-type soldering irons. Discrete collars, with two or four sides, are available for specific heating requirements. Soldering collars are designed primarily for use with multiple lead components, such as integrated circuits; however, they are also effective for the removal of rectangular and cylindrical components. Collars are useful when removing components that have been attached with adhesive. After the solder joints have reached liquidus the collar can be used to twist the component, which breaks the adhesive bond. Soldering collars are mainly intended for component removal.

A problem does occur with four-sided components, such as the PLCC, because it is difficult for the collar to contact all of the leads at the same time. If the lead does not contact the collar heat transfer does not occur, resulting in solder joints that do not achieve liquidus. This is not necessarily the fault of the collar. On large integrated circuit packages all of the leads are not usually in the same plane of reference, which makes it impossible for the collar to contact all of the leads simultaneously. This condition is dangerous because the operator may believe all of the solder joints have reflowed when they have not. When the operator attempts to remove the component, lands that are still soldered to leads will be pulled from the PCB.

Soldering tips and collars require good proactive maintenance. They need to be kept clean, and in some cases properly tinned with solder. Frequent replacement may also be required, especially when using small tips. See Figure 11.2 for examples of various attachments.

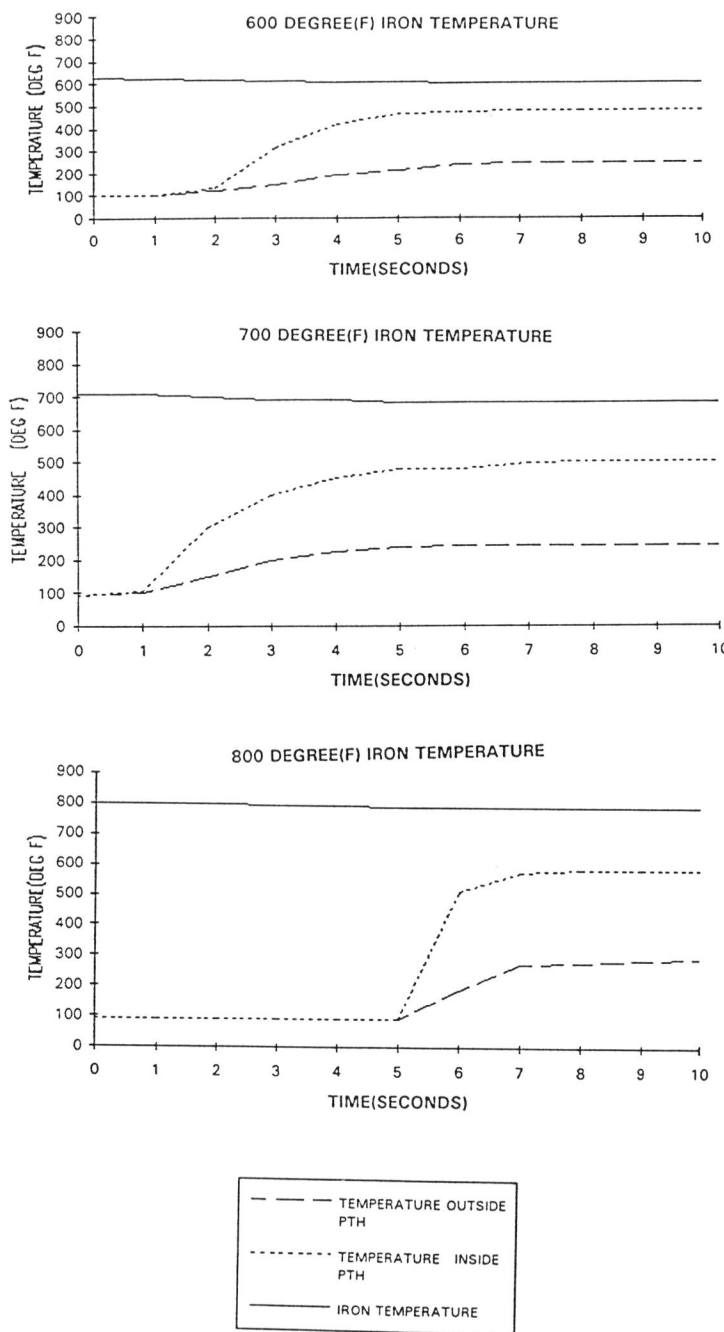

FIGURE 11.1. Relationship Between Soldering Iron Tip Temperature and the PTH Temperature.

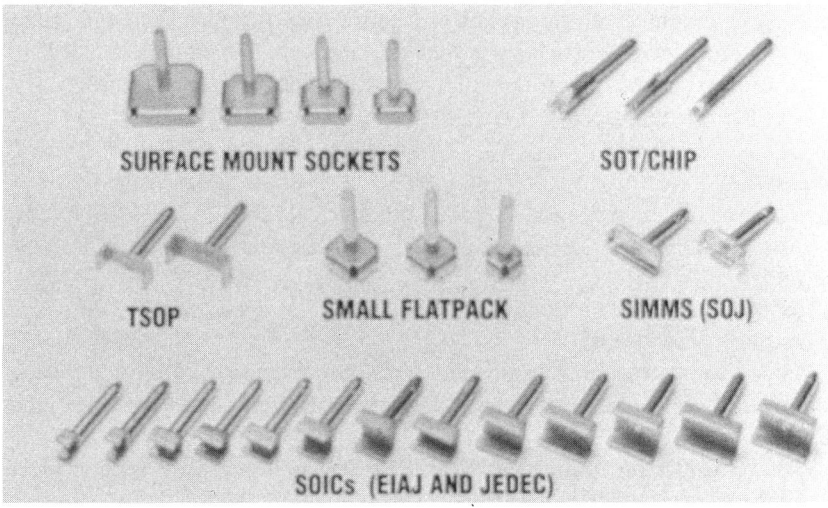

FIGURE 11.2. Soldering Iron Tip Attachments. (Courtesy of Pace, Inc.)

11.2.1 Contact Soldering Systems

There are different types of contact soldering equipment. They can be categorized, in order of lowest cost to highest cost, as constant temperature, restricted temperature, and controlled temperature systems. Selection of a particular system depends on how delicate a task it is required to perform. Keep in mind that surface mount applications require considerably less heat than through-hole applications.

Constant temperature systems have a continuous, constant output; in other words, they deliver heat continuously. For surface mount applications a system should operate within a temperature range of 335°C to 365°C (635°F to 689°F).

Restricted temperature systems have a temperature restricting capability that does a better job of keeping the system temperature within a preferred range. These systems do not continuously deliver heat, which is what prevents them from overheating; however, their heat renewal can be slow. This problem sometimes causes operators to set the temperature higher than desired in an effort to solder faster. The operating temperature range for surface mount applications is about 285°C to 315°C (545°F to 599°F).

Controlled temperature systems have a high output capability, which is turned on as required. These systems, like temperature restricted systems, do not continuously deliver heat; however, their response time and temperature control are much better than restricted temperature systems. The operating temperature range for surface mount applications is also about 285°C to 315°C (545°F to 599°F), but the tolerance capability is much better, usually less than 10°C (50°F).

Attributes
For removal and replacement of simple individual components a good temperature-controlled soldering iron is the quickest, easiest, and lowest-cost method.

Components attached with adhesive can be removed quickly and easily with collar-type systems.

It takes very little time to properly train an operator to use a soldering iron. A skilled operator can make acceptable solder joints with minimal effort.

Contact soldering equipment is generally low in cost. There are countless suppliers, so availability is excellent.

Concerns
With hand-held equipment the operator is always a variable. If the operator uses an excessive tip or collar temperature, above 370°C (698°F), the operator might unknowingly cause quality and reliability problems. Systems that do not restrict or control the tip or collar temperature are susceptible to temperature spikes that can drive the tip or collar temperature above the preferred range.

The heated tip or collar must make direct contact with the solder joint and/or component lead.

Contact soldering irons can thermal shock some components, causing them to crack. Ceramic components are the main concern, especially multilayer capacitors.

Operators may tend to use the tip or collar as a pry bar. This can result in excessive pressure being applied to the component and PCB, causing damage to the component and/or PCB.

Component-specific collars are required.

11.3 HOT GAS SOLDERING

Hot gas soldering is accomplished by heating air, or an inert gas such as nitrogen, and directing it at the solder joint or joints with a nozzle. Hot gas equipment ranges from simple hand-held units designed to heat a single solder joint to complex automated units designed to heat multiple solder joints. Automated systems are usually designed to remove and replace the component, as well as heat the solder joints. This is especially helpful with

FIGURE 11.3. Hot Gas System Example.

fine-pitch components. Hand-held systems are used to remove and replace rectangular, cylindrical, and SOT-type components. The larger semiautomatic and automatic systems are used to remove and replace integrated circuit packages, such as the SOIC, PLCC, and fine-pitch components.

Hot gas systems heat the entire component uniformly. This makes it a good choice for removing ceramic components, since it eliminates the localized heat stress that can occur with single-tip contact soldering systems. In fact, hot gas is the preferred method for replacing *all* components because it heats the component uniformly. The heated gas temperature is usually between 300°C and 400°C (572°F and 752°F). The solder joints should be heated to a temperature between 200°C and 220°C (392°F and 428°F), depending on the solder alloy used. See Chapters 5 and 7 for additional information on reflow temperatures. The time required to remove a component decreases as the hot gas volume increases. Some large components, such as the PLCC68, may require up to 60 seconds of heating before they can be removed. Most systems designed for large component removal have timers that can be used. An example is shown in Figure 11.3.

Nozzle design is critical. The nozzle must direct the hot gas to the solder joints and, in the case of integrated circuits, away from the component body. The nozzle must also be durable. Nozzles for large components are usually complex and quite expensive. Proper care, in the form of good proactive maintenance, is essential. Nozzles should be cleaned to avoid flux build-up and properly stored to prevent damage.

Attributes

Overall, hot gas is the best method available for the removal and replacement of surface mount components. The temperature and heating rates are controllable, repeatable, and predictable. The inefficiency of gas as a heat transfer medium works in its favor. Component and PCB thermal shock is avoided because of the slow heating rate.

Operator training is not too difficult.

There are numerous suppliers of hot gas systems, so equipment availability is good.

Concerns

Hot gas systems range in price from moderately expensive to very expensive.

Component-specific nozzles are required. Nozzles can become quite expensive.

With hand-held equipment the operator is always a variable. Semiautomatic and automatic systems are complex and require a high level of operator skill to operate.

11.4 INFRARED SOLDERING

Infrared (IR) soldering systems, pictured in Figure 11.4, have emerged as a viable alternative for surface mount repair in the last couple of years. Their main value, other than speed, is the ability to rework components with multiple leads, especially high lead count and fine-pitch components. Infrared systems operate by focusing IR light onto the solder joints to achieve liquidus. In some cases only a small number of adjustable lens attachments or nozzles are required to cover a wide range of components. Many systems also utilize a bottom-side IR or convection heater to preheat the PCB. Most systems have temperature settings up to 300°C (572°F). The solder joints should be heated to a temperature between 200°C and 220°C (392°F and

Rework   209

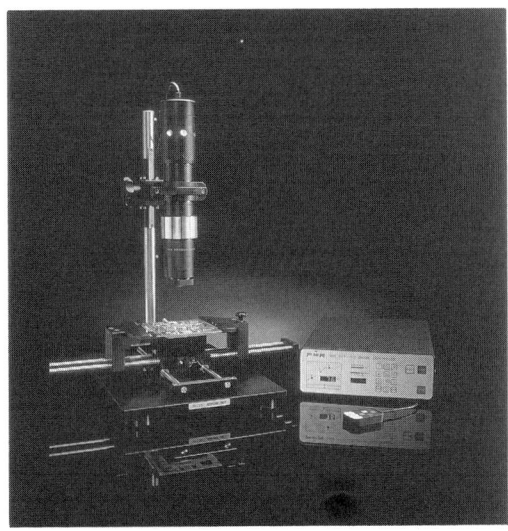

FIGURE 11.4.   Focused Infrared Soldering System Example. (Courtesy of ISMECA USA, Inc.)

428°F), depending on the solder alloy used. For a complete review of IR energy see Chapter 8.

As is the case with hot gas soldering systems, lens or nozzle design is critical. The lens or nozzle must direct the IR light to the solder joints and, in the case of integrated circuits, restrict heating of the component body. These attachments, which tend to be fragile, are usually complex and quite expensive. Proper care and good proactive maintenance are essential. Careful storage is required to prevent damage.

Attributes

The temperature and heating rates are controllable, repeatable, and predictable.

Several lens or nozzle attachments may cover a wide range of surface mount components.

Infrared systems are faster than hot gas systems.

Concerns

Infrared systems range in price from moderately expensive to very expensive.

Component-specific nozzles are sometimes required. Lens and nozzle attachments can become quite expensive.

Infrared systems are complex and require a high level of operator skill to operate.

Availability is limited to a few suppliers.

11.5 GENERAL REQUIREMENTS

11.5.1 Heat Stress

Rework time and thermal shock can be reduced by preheating the PCB for three to five minutes at approximately 100°C (212°F). Avoid preheating the PCB above its glass transition temperature (Tg), usually 125°C (257°F), for more than two minutes. The entire PCB can be heated, or just the area that is to be reworked. Batch convection ovens or hot plates can be used to preheat the entire PCB. Some rework systems have the ability to preheat localized areas of the PCB.

Avoid reworking any component location more than two times. Repeated localized heating of the PCB can cause immediate damage (such as lifted lands) or long-term reliability concerns (such as weakened plated through-holes that will fail after continued cycling in the product).

11.5.2 Flux and Solder

During rework solder can be applied to the solder joint in two forms: wire/flux core solder (commonly called core solder) or solder paste dispensed from a syringe. It is usually best to use the same solder alloy for rework that is being used for reflow and wave soldering. See Chapter 5 for information on fluxes, solder alloys, and solder paste.

Core solder contains flux as well as the solder alloy. Using core solder frees the operator from having to apply flux as well as solder during the rework process. The flux is placed at the center, or core, of the wire. The most common flux in use is an RMA; however, R, RA, and water-soluble fluxes are also available. Wire diameter is important. Surface mount rework requires a small diameter wire in the 0.5mm to 0.75 mm (0.020" to 0.030") range, while through-hole rework requires a large diameter wire in the 1.2 mm to 1.5mm (0.047" to 0.059") range.

It is usually a good idea to have a small bottle of liquid flux available for reworking solder joints that do not require additional solder. Several different sizes of tweezers are valuable for handling components.

11.5.3 Jumper Wires

Adding jumper wires to surface mount components is a challenge. The solder joints and leads are much smaller, which leaves very little area to attach the wire to. A 30 AWG normally works well with surface mount components. When adding jumper wires, the first choice is to solder the wire to a plated through-hole. When this is not feasible, and the wire needs to be attached to a lead, the end of the wire should be formed into a hook. The wire hook is then hooked around the lead and soldered in place. Use a small amount of solder to attach the wire to the lead. An excessive amount of solder will stiffen the lead, decreasing its compliance. If the wire is to be attached to a termination it should be soldered, in a lap joint configuration, to the face of the component or the top of the land.

11.5.4 Adhesive

When a component that was attached with adhesive is removed any excess adhesive on the PCB should be removed as well. Adhesive left on the PCB could interfere with the proper attachment of a new component. Since the adhesive was only used to hold the component in place during wave soldering, it is not necessary to use adhesive when a component is being replaced by hand.

11.5.5 Land Preparation

After the component has been removed the lands on the PCB should be inspected for damage and surface condition. Serious damage has occurred if a land has been lifted. There are methods available to replace lands, but their long-term reliability is a concern. If a rectangular, cylindrical, or SOT-type component has a land lifted, the component can be attached to the PCB with a surface mount adhesive. A jumper wire can then be attached to the termination or lead, and then to another point on the PCB. A similar approach can be used with multiple lead components (SOIC, PLCC, etc.); but if only one land is damaged the other soldered leads will hold the component in place, so an adhesive is not necessary to hold the component.

Lands should also be inspected for solder quality. If the solder forms an irregular surface on the land, it may be necessary to reflow it with a rework tool to achieve a smooth, uniform surface. A smooth, uniform surface makes it much easier to replace the component, especially multiple lead components. It may also be necessary to remove the old solder and replace it with fresh solder. This can be done by removing the old solder with a braided solder wire and adding new solder using a core solder or solder paste.

11.6 WORKMANSHIP STANDARDS

Workmanship requirements and standards vary, sometimes considerably, from company to company. A good standard reference document is ANSI/J-STD-0001, "Requirements for Soldered Electrical and Electronic Assemblies," which is available from IPC (see Appendix A for ordering information). There are also numerous workmanship standards available from consultants and companies. As noted earlier, good workmanship standards are a must. If no standard is established, operators will develop their own, and each one will be different. This means acceptable solder joints will be rejected and defective solder joints will be accepted. Many workmanship standards have three classifications: preferred, acceptable, and unacceptable. Only two classifications are necessary, acceptable and unacceptable. The solder joint is either good or bad. Having a preferred condition only confuses the issue. Operators will often rework acceptable solder joints so they meet the preferred requirement. In addition to being a waste of time and money, it causes reliability problems. A reworked solder joint is never as good as the original, if it was acceptable to begin with. When a solder joint is borderline, *do not* rework it!

REFERENCES
1. ANSI/J-STD-0001. "Requirements for Soldered Electrical and Electronic Assemblies." IPC, Lincolnwood, IL, April 1992.
2. Bergenthal, Jim. "Surface Mount Technology Repair, Touch Up and Hand Soldering: Can these be controlled?" Kemet Electronics Corporation, January 1989.
3. Dow, Steve and Helton, David. "The Use of Collimated Infrared Light." *Circuits Assembly*, November 1990, pp. 28–30.
4. Kiernan, Terry. "SMT Rework Roundup." *Circuits Manufacturing*, August 1990, pp. 25–36.
5. Manko, Howard H. *Soldering Handbook for Printed Circuits and Surface Mounting*. New York: Van Nostrand Reinhold, 1986.
6. ———. *Solders and Soldering,* New York: McGraw-Hill, 1979.
7. Morency, Daniel. "A Discussion of SMT Solderability Issues and Relationships to Lead Finish," *Surface Mount Technology*, June 1991, pp. 30–34.
8. Morris, Barry. "The Practicalities of Surface Mount Rework." *Printed Circuit Assembly*, November 1989, pp. 22–28.
9. Prasad, Ray P. *Surface Mount Technology—Principles and Practice*. New York: Van Nostrand Reinhold, 1989.
10. Wolverton, Mike. "Component Solderability," *Circuits Assembly*, March 1991, pp. 34–42.

12
Manufacturing Operations

GLOSSARY

Automated Material Handling   The automatic transfer of product between manufacturing operations using conveyors and other automated equipment.

Batch Flow Manufacturing (BFM)   A manufacturing operation that processes product in groups.

Continuous-Flow Manufacturing (CFM)   A manufacturing operation that processes product one at a time.

Flow Lines   A manufacturing configuration that places equipment in the order of assembly.

Kanban (pronounced kahn-bahn)   A Japanese term that basically means "a visible record." Japanese companies use a kanban card to signal the need for more material to be delivered to a certain point. The Japanese use of the kanban card differs from the American shop traveler because the kanban card is a pull system, whereas the shop traveler is a push system.

Lot   A manufacturing unit that contains any number of the same product. The product will remain together in this unit through the manufacturing process.

Lot Size   The amount of product in a lot.

Work Cell   A manufacturing configuration that places all equipment with the same function together.

12.0 INTRODUCTION

The manufacturing world has gone through a major revolution in the past decade. It is very important to understand the changes that have occurred. Buying equipment and implementing manufacturing processes are the beginning of the production process. Once these are complete the operation must be carefully managed to be completely successful.

Manufacturing operations is a very broad subject. This chapter does not explain manufacturing operations in detail. That would require another book. The intent of this chapter is to give the reader some important ideas to consider, and possibly research further. For additional information read the books listed in the reference section of this chapter.

This chapter will review such issues as lot-size and set-up considerations, process flow, automated material handling, and equipment maintenance.

12.1 LOT-SIZE CONSIDERATIONS

One of the first issues to consider when developing a manufacturing strategy is lot size. Twenty years ago this surely would have been based on an economic lot size; however, the Japanese have forever changed the lot-size concept. Modern manufacturing operations must be very flexible, and they must be able to cope with constant change.

12.1.1 Economic Lot Size

An economic lot size (ELS) is a compromise between inventory cost and set-up cost. The ELS formula was developed around 1915, and it has been used extensively since then. Excessive set-up time is the basic premise for using an ELS. In other words, if set-up takes a long time to accomplish it should be avoided as much as possible. To minimize changeover and set-up, larger lots are favored. Inventory cost and set-up cost are the apparent costs to be concerned about. However, quality, motivation, and productivity are also affected by the lot size. It is time to abandon some ELS concepts and utilize new concepts based on a minimum lot size and set-up time reduction.

The graph shown in Figure 12.1 illustrates how lot size is affected by the frequency of set-up and the set-up time. The lot size in turn affects inventory costs. An excessive set-up time mandates infrequent set-up, which results in a large lot size being produced. The large lot size requires additional material, which increases inventory costs.

12.1.2 Minimum Lot Size

The minimum lot size (MLS) concept strives to decrease the lot size by radically reducing the set-up time, which encourages more frequent set-up. The Japanese developed and implemented the MLS concept in the late 60s and early 70s. They are still pursuing and refining the concept with vigor. With proper attention both set-up time and set-up cost can be reduced dramatically. In addition to set-up and inventory cost savings, there will be improvements in quality, motivation, and productivity.

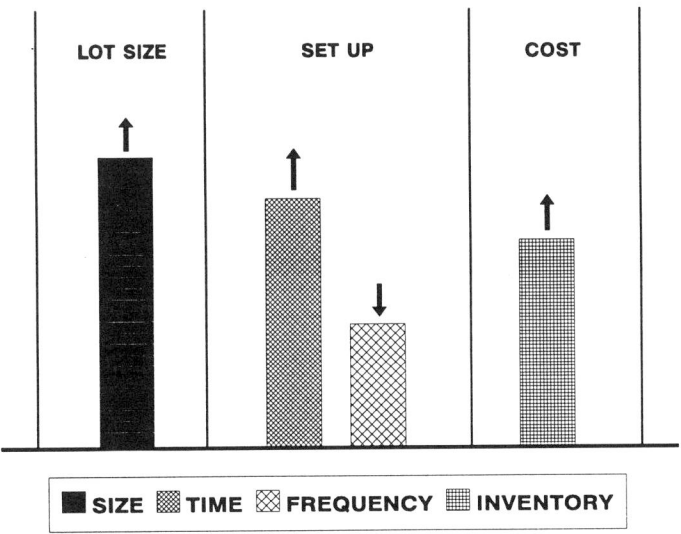

FIGURE 12.1. Economic Lot Size Illustration.

The graph in Figure 12.2 again illustrates how lot size is affected by the frequency of set-up and the set-up time. In this case, however, a low set-up time has been accomplished. Notice that the low set-up time allows the frequency of set-up to increase, which decreases the lot size that must be produced. A smaller lot size decreases material requirements, and as a result reduces inventory costs.

12.1.3 ELS and MLS Comparison

A simple formula can be used to determine lot size requirements. It can also be used to show the difference between an economic lot size and a minimum lot size, and the effect that set-up time will have on both.

The formula takes into account four items: annual PCA production, set-up cost, PCA cost, and inventory carrying charge rate. Annual PCA production is the total number of PCAs of that type produced in one year. Set-up cost is the total cost to set up the line to produce that PCA. The PCA cost is the total cost of the PCA, including material, labor, and overhead. The inventory carrying cost, expressed as a percentage of the value of the inventory, includes such costs as management, scrap, storage, interest, and taxes. The inventory

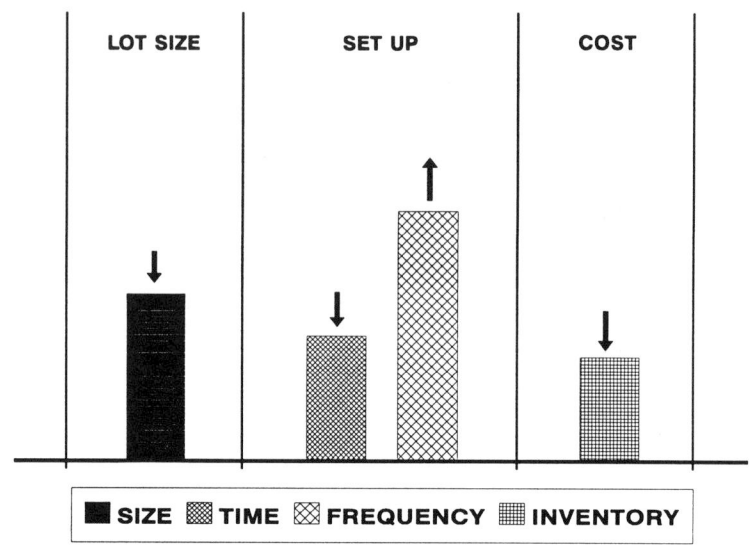

FIGURE 12.2. Minimum Lot Size Illustration.

carrying cost will vary from approximately 15% to 25% of the value of the inventory.

$$\text{Lot Size} = \sqrt{\frac{2 \times (\text{Annual PCA Production}) \times (\text{Set-up Cost})}{(\text{PCA Cost}) \times (\text{Inventory Carrying Cost})}}$$

EXAMPLE

Determine the lot size based on the following information:

Item	ELS	MLS
Annual PCA production:	50,000	50,000
Set-up cost:	$720	$180
PCA cost:	$300	$300
Inventory carrying cost:	15%	15%
LOT SIZE:	1,265	632

Notice that because the set-up time decreased the set-up cost decreased by 75% ($720 versus $180). The lot size was reduced by 50% as a result of the decreased set-up time and cost. This is only an example but it does illustrate what can happen with a significant reduction in set-up time and cost.

12.2 SET-UP CONSIDERATIONS

12.2.1 Set-up Time Reduction

Japanese companies were the first to realize the potential of decreasing set-up time and reducing lot size. World class manufacturers, such as Toyota with their Single-Minute-Exchange-of-Die (SMED) program, have focused on and fine tuned their set-up procedure to decrease set-up time and improve the repeatability of the manufacturing process. To achieve a short set-up time requires a continuous, dedicated effort to optimize the product and process for an efficient set-up. Set-up must be viewed as proactive rather than reactive. In other words, set-up must be planned for, whether before a new job is started or after it is running. For example, when changing over to a new job at screen printing have the new stencil available and waiting near the screen printer. Do not wait until the last minute to start looking for the stencil. And when a job is already running on a placement system have extra reels of components loaded on a feeder so when the system runs out of components changeover can occur quickly. Too often operators will let the system run out of components before they bother to obtain more.

To illustrate this point let's look at a successful example of efficient set-up in another industry. A vital aspect of Indy-type auto racing is the ability to make a quick and successful 'pit stop' to add fuel, change tires, etc. Notice that each car is designed so that parts can be changed quickly and correctly. In addition, the 'pit crew' is trained to add fuel and change parts properly in the shortest possible time. If they accomplish their goal satisfactorily (a short set-up time) they give their team a competitive edge. Rarely does a team win the race if their pit crew performs poorly.

This lesson can be applied directly to a manufacturing operation. Too often set-up is viewed as something that "just happens." Do you think an Indy 'pit stop' occurs with no planning, training, or practice? Set-up in a manufacturing operation must be viewed in the same fashion. Production equipment must be designed, and continually redesigned if necessary, for a quick and reliable set-up. The product (PCA) should also be designed to aid set-up as much as possible. As in the example, the team of operators responsible for set-up must be very well trained. If they can achieve a quick, efficient set-up they give their company a competitive edge.

12.2.2. Set-up Aids

An important element in the selection and design of new equipment is ease of set-up. For example, design equipment so that all adjustments can be made quickly by hand with no tools. This means eliminate the use of items that require screwdrivers, wrenches, and other hand tools. Instead, use quick-release clamps, threaded knobs, magnets, Velcro—anything that will remove the need to find and use a hand tool. The PCA can also be designed to aid the set-up process. Use one standard width for all PCBs; add break-off tabs if necessary. Standardize PCB tooling hole location and diameter. Determine common use components, and require designers to select from that list.

Training is the one area that must not be overlooked. The set-up procedure must first be developed and documented. Then operators must be trained and certified. Practice is very important. The set-up process must become routine.

There are many benefits to set-up time reduction. The most obvious is a decrease in set-up cost. As noted earlier this will have a positive effect on lot size requirements. Spending less time on set-up also decreases the cycle time, which means that more time can be spent producing product.

12.2.3 Set-up Guidelines

Set-up should not be a difficult process. The best way to be successful at set-up is to keep things simple. Use the following guidelines to help implement a quick and simple set-up process.

- Select equipment that is easy to set up
- Redesign old equipment to improve set-up
- Design new equipment for ease of set-up
- Use quick-release tooling, such as clamps
- Design the PCA to aid the set-up process
- Use proactive rather than reactive set-up
- Develop and document set-up procedures
- Train operators, then practice set-up
- Keep all tools well maintained and available
- Continuously analyze and refine the set-up process

12.3 PROCESS FLOW

12.3.1. Batch Flow Manufacturing

Batch flow manufacturing (BFM) produces product in a cluster. A group of PCBs is processed at one operation and then moved to the next operation.

The group of PCBs stays together through the entire manufacturing process. As an example, ten PCBs are processed at the screen printing operation. They are then moved together to the placement operation, and so on.

Low-volume, high-mix manufacturing operations have traditionally been based on batch flow manufacturing concepts. Batch flow manufacturing can also be considered a push system because material movement is dependent on the operator moving (pushing) the PCBs through the manufacturing operation. There is usually not an efficient, continuous flow of PCBs through the entire manufacturing process.

12.3.2 Continuous Flow Manufacturing

In contrast to batch flow manufacturing, continuous flow manufacturing (CFM) produces product one at a time, and then immediately moves it to the next operation. High-volume, low-mix manufacturing operations have traditionally been based on continuous flow manufacturing concepts. Concern over set-up time and lot size has prevented the acceptance of continuous flow manufacturing for low-volume, high-mix production. However, as noted earlier, with proper attention both set-up time and lot size can be reduced dramatically. Continuous flow manufacturing can also be considered a pull system because each PCB is efficiently pulled to the next manufacturing process as required.

One of the keys to continuous flow manufacturing is to view the manufacturing operation as one continuous system, rather than as a collection of individual cells. Contrary to popular belief, a continuous flow manufacturing operation does not necessarily require the use of an automated material handling system (conveyors). However, some method is required to control, or pace, the flow of product between operations. If an automated material handling system is used this can be accomplished with the use of conveyors and optical sensors. On the simpler side, product can be transferred manually with the aid of control points. One method is the use of a "kanban square" as a control point between operations. This is really a very simple concept. A small table, with one square drawn at the center, is placed between operations (for example, between placement and reflow). When the PCA has completed the placement operation it is positioned on the table. If the reflow operation is ready to accept the PCA it is removed from the table and placed in the reflow system. If the reflow operation is not ready the PCA remains on the table. The key element here is that the placement operation can not process another PCA until the table (kanban square) is open. This simple method will do a very good job of controlling and pacing a manufacturing line.

12.3.3 Work Cells

The work cell concept groups similar processes and equipment together in the same area. For example, all screen printers are in one group, all placement equipment is in another group, and so on. See Figure 12.3. This has been a common layout method for surface mount manufacturing. There are some obvious concerns about this method, including excessive material handling, poor coordination with the previous and subsequent process, and the need for large lot sizes. A work cell tends to become an invisible island that is separated from the other manufacturing operations. This can and will hinder continuous improvement.

12.3.4 Flow Lines

The flow line concept places discrete but related processes and equipment in a serial format. The purpose is to get subsequent processes as close together as possible. This forces related operations to work together, which will improve communication and solve problems much faster. In addition, less material handling is required and lot sizes can be smaller.

Flow lines can be set up in several ways. Common configurations are linear, L-shaped, and U-shaped. See Figures 12.4, 12.5, and 12.6 for examples. Line configuration is usually related in some way to the shape of the building. American companies tend to favor linear or L-shaped

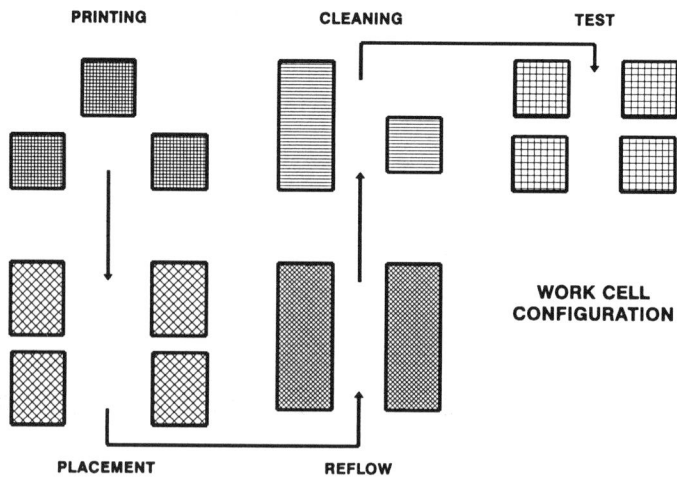

FIGURE 12.3. An Example of the Work Cell Concept.

Manufacturing Operations 221

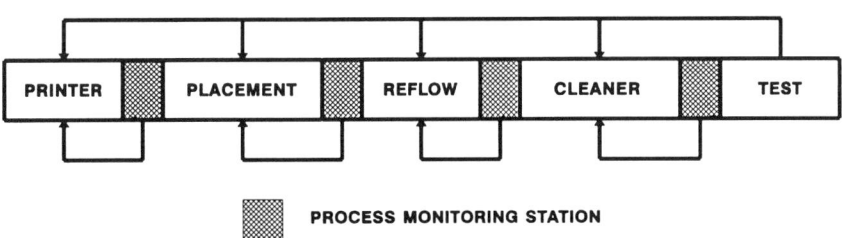

FIGURE 12.4. An Example of the Linear Flow Line Concept.

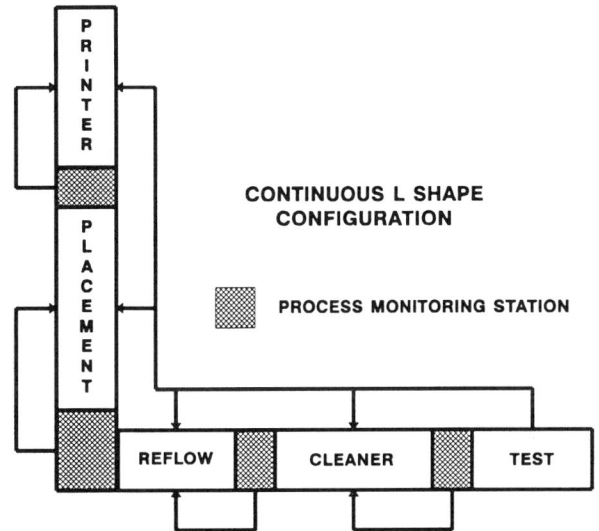

FIGURE 12.5. An Example of the L-Shaped Flow Line Concept.

configurations, while Japanese companies often use a U-shaped configuration. The focus in most American companies is line balance (which favors the linear configuration), while Japanese companies focus more on flexibility (which favors the U-shaped configuration). U-shaped configurations allow one operator to handle tasks on both sides of the manufacturing line.

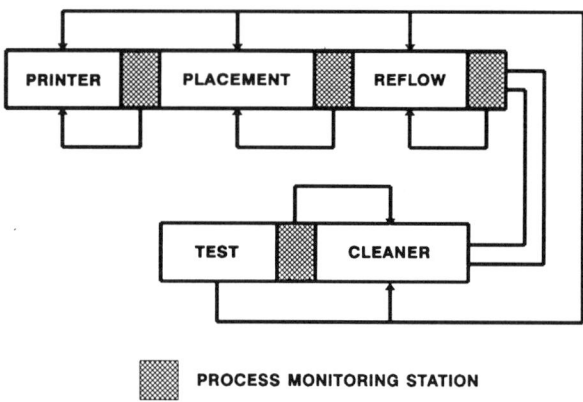

FIGURE 12.6. An Example of the U-Shaped Flow Line Concept.

12.4 AUTOMATED MATERIAL HANDLING

12.4.1 Material Handling Equipment

Automated material handling equipment is used to load, transfer, store or buffer, and unload product from a manufacturing line. Too often these systems are much larger and more complex than they should be. The key element to remember is that the sole purpose of the material handling equipment is to safely transfer product between operations. There is some need to control, pace, and even buffer product, but do not get carried away, especially with buffering (remember the kanban concept).

The main element in all automated material handling systems is the edge-hold conveyor, which is used to transfer the PCB. See Figure 12.7. There are three common conveyor types, based on the method they use to carry the PCB. The oldest type is the chain conveyor. A later development is the "power and free" conveyor, which is a chain conveyor with injection-molded wheels that rotate independent of the chain. The latest development is the belt conveyor, which uses a cloth or rubber belt in place of the chain. Chain conveyors usually use a mechanical block to stop the PCB. The chain continues to move while the PCB is at the block, which will vibrate the PCB. For surface mount applications it is more common to use the belt conveyor, with an optical sensor to stop the PCB. With this application the belt does not move and the PCB is not subjected to any vibration. The belt/optical sensor method is more expensive than the chain/mechanical block method because the belt conveyor must be purchased in short, individual sections.

Manufacturing Operations   223

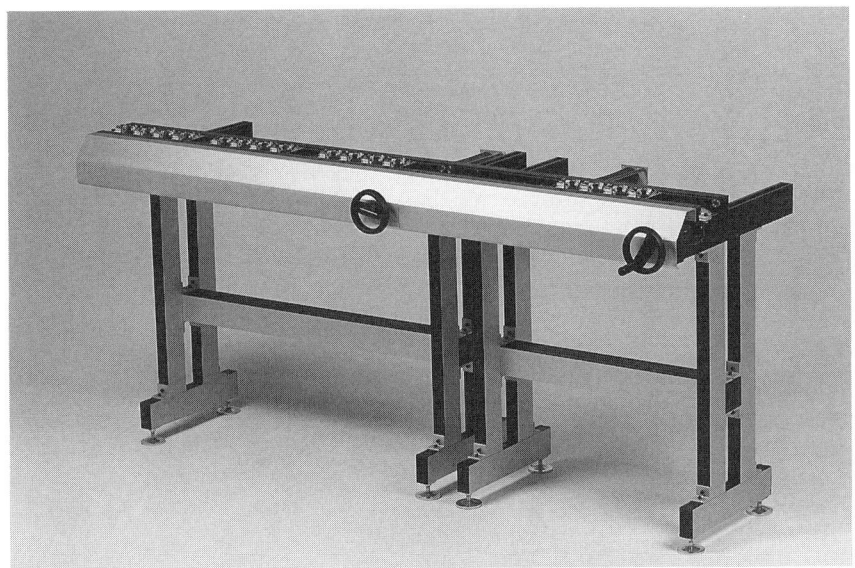

FIGURE 12.7. A Common Edge Hold Conveyor.

A flat belt conveyor, made from a solid piece of cloth or a stainless steel mesh, is also used in limited applications. This can only be used in situations where alignment of the PCB is not important, such as transferring the PCB between two belt-driven machines (for example, reflow system to cleaning system).

Other types of material handling equipment include the following:

Destacker—transfers bare PCBs from a stack into the line.

Magazine Unloader—transfers bare or partially completed PCBs from a slotted magazine into the line.

Turntable—rotates a PCB 90°. See Figure 12.8.

Shuttle—transfers a PCB between parallel lines.

Inverter—inverts a PCB so the opposite side can be processed.

Lift Gate—section of conveyor that lifts up into a vertical position to allow passage to the other side of the line.

Brush Aligner—orients a misaligned PCB from a flat or mesh belt conveyor to an edge handling conveyor. See Figure 12.9.

Vertical Buffer—buffers PCBs between operations using last-in-first-out (LIFO) or first-in-first-out (FIFO) operation. FIFO is preferred for surface mount applications. See Figure 12.10.

Loader—transfers a PCB from the line into a slotted magazine.

FIGURE 12.8. A Common Turntable.

FIGURE 12.9. A Common Brush Aligner.

12.4.2 Automation and People

People have an extremely important, but very often overlooked, role in an automated environment. Too often automation is considered a large-scale replacement for people. This is simply not true. Machines, whether simple or complex, require the attention and dedication of highly skilled and motivated people. Automate to prevent variation in a process, eliminate mundane operations, and remove safety hazards. Do not automate simply to reduce direct labor cost.

12.4.3 Automation Guidelines

- Use automation wisely—do not over-automate
- Automate to reduce variation in the process
- Automate to eliminate safety hazards
- Do not automate solely to eliminate direct labor
- Automation should complement people not replace them
- Design the product for an automated environment

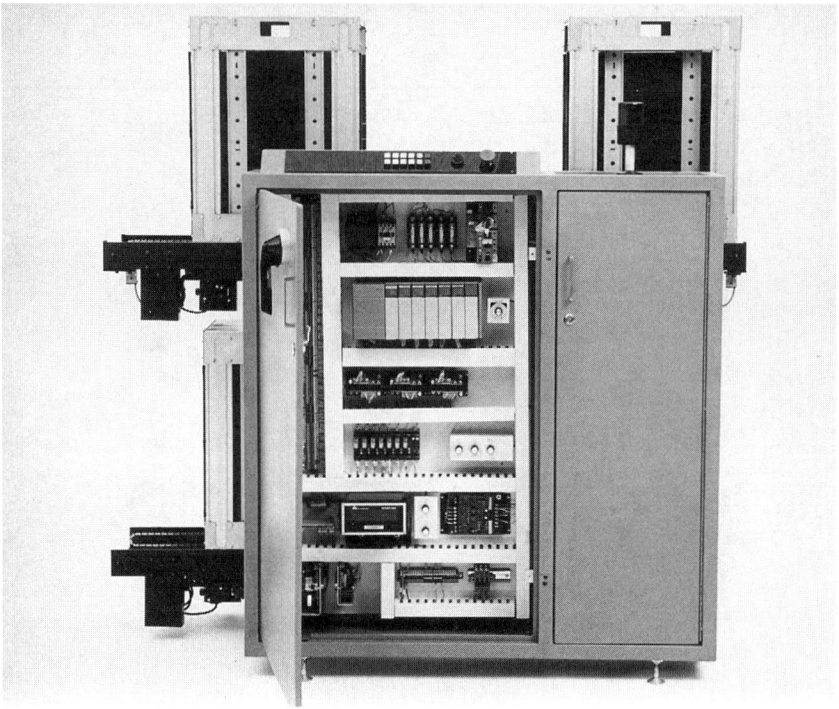

FIGURE 12.10. A Common Vertical FIFO Buffer.

- Use continuous flow manufacturing concepts
- Use simple, flexible automation
- Keep operations close together
- Keep conveyor lines short and simple
- Use conveyors to transfer product, not store it

12.5 EQUIPMENT MAINTENANCE

12.5.1 Proactive Maintenance

The best way to ensure that a manufacturing operation continues to produce product is to use proactive rather than reactive maintenance. Proactive maintenance is done before the equipment fails; reactive maintenance is done after the equipment fails. Proactive maintenance is planned; reactive maintenance is unplanned. This sounds like a simple concept, but the truth is many companies do not do an adequate job of proactive, or preventative, maintenance. A lack of management commitment, desire, and support is usually the problem. Management must be made aware of the value of proactive maintenance. With proper leadership and support from management, operators can gain a feeling of ownership and responsibility for their equipment. Once this support is achieved, the operator will continue to keep the equipment clean and well maintained.

12.5.2 Maintenance Guidelines

- Use proactive rather than reactive maintenance
- Develop and document maintenance procedures
- Establish goals for equipment availability
- Train operators to maintain and repair equipment
- Respect and use operator ideas and input
- Keep tools well maintained and available
- Implement a system to record equipment problems
- Analyze failure reports to determine cause of problem
- Track parts usage and reorder when necessary

12.5.3 Manufacturing Productivity Factors

Several factors can help to understand the manufacturing operation; they include equipment availability, utilization, and productivity. Availability is the amount of time, expressed as a percent of the total available manufacturing hours, the equipment is available to manufacturing. Utilization is the amount of time, expressed as a percent of the total available hours, the equipment is used out of the time it is available. Productivity is the amount of time, expressed in percent, the equipment is used to produce product as shown in the following formulas. These formulas can be altered to include other factors, but the basic thought process would be the same.

$$\text{Availability} = \frac{\text{Available Hours}}{\text{Total Hours}}$$

$$\text{Utilization} = \frac{\text{Used Hours}}{\text{Available Hours}}$$

$$\text{Productivity} = (\text{Availability}) \times (\text{Utilization})$$

Figure 12.11 shows an example of a type of chart that could be used to display information that is important to the manufacturing operation.

228 Applied Surface Mount Assembly

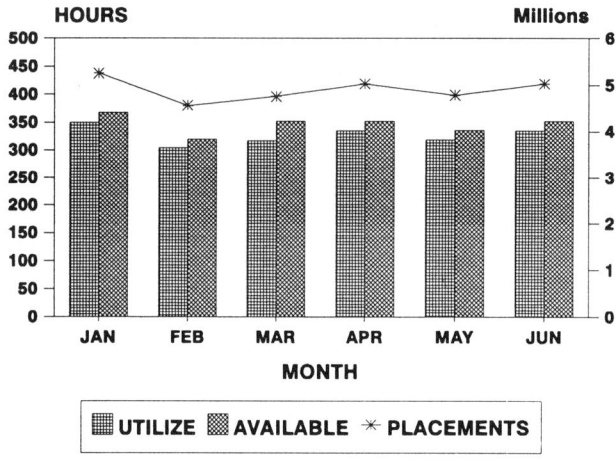

FIGURE 12.11.   An Example of a Equipment Productivity Chart.

REFERENCES
1. Moore, Franklin G. *Production Control.* New York: McGraw-Hill, 1969.
2. Rowland, Robert J. "Continuous Flow Manufacturing at Teradyne." *Circuits Assembly*, February, 1991, pp. 50–52.
3. ———. "Designing Surface Mount Assemblies for Automation." *Printed Circuit Design*, November 1986, pp. 30–37.
4. Schonberger, Richard J. *Japanese Manufacturing Techniques.* New York: Macmillan, 1982.
5. ———. *World Class Manufacturing.* New York: Macmillan, 1986.
6. Shingo, Shigeo. *Study of Toyota Production System from Industrial Engineering Viewpoint.* Japan Management Association, Tokyo, 1981.

Appendix A
Documentation

Numerous documents, standards, and specifications have been developed by organizations that supply and/or use electronic components and hardware. This appendix contains the names of the organizations in the United States and a list of the documents, standards, and specifications that they publish.

ORGANIZATIONS

EIA	Electronic Industries Association
	2001 Pennsylvania Avenue, NW
	Washington, DC 20006-1813
IEC	International Electrotechnical Commission
	3 Rue de Varembe
	1211 Geneva 20, Switzerland
	Obtain IEC Documents from:
	American National Standards Institute (ANSI)
	11 West 42nd Street
	New York, NY 10036
IPC	Institute for Interconnecting and Packaging Electronic Circuits
	7380 North Lincoln Avenue
	Lincolnwood, IL 60646
JEDEC	Joint Electron Device Engineering Council of the EIA
	Obtain Documents from EIA
MIL	Military
	Standardization Documents
	Order Desk, Building 4D
	700 Robins Avenue
	Philadelphia, PA 19111-5094

SMEMA   Surface Mount Equipment Manufacturers Association
         4113 Barberry Drive
         Lafayette Hill, PA 19444

Components, Passive

EIA-575	Resistors: Rectangular, Surface Mount, General Purpose
EIA-576	Resistors: Rectangular, Surface Mount, Precision
EIA/IS-34	Leaded Surface Mount Resistor Networks: Fixed Film
EIA-469-B	Standard Test Method for Destructive Analysis of High Reliability Ceramic Monolithic Capacitors
EIA-CB-11	Guidelines for the Surface Mounting of Multilayer Ceramic Chip Capacitors
EIA/IS-28	Fixed Tantalum Chip Capacitor Style 1 Protected—Standard Capacitance Range
EIA/IS-29	Fixed Tantalum Chip Capacitor Style 1 Protected—Extended Capacitance Range
EIA/IS-36	Chip Capacitors, Multilayer (Ceramic Dielectric)
EIA/IS-37	Multiple High Voltage Capacitors
IEC-384-3	Sectional Specification, Tantalum Chip Capacitors
IEC-384-10	Sectional Specification, Fixed Multilayer Ceramic Chip Capacitors
IECQ Draft	Blank Detail Specification, Fixed Multilayer Ceramic Chip Capacitors
IECQ-PQC-31	Sectional Specification, Fixed Tantalum Chip Capacitors with Solid Electrolyte
IECQ-PQC-32	Blank Detail Specification, Fixed Tantalum Chip Capacitor

Components, Active

EIA-JEP-95	JEDEC Registered and Standard Outlines for Semiconductor Devices
EIA-JESD21-C	Configurations for Solid State Memories
EIA-JESD22-B	Test Methods and Procedures for Solid State Devices Used in Transportation/Automotive Applications
EIA-JESD-11	Chip Carrier Pinouts Standardized for CMOS 4000, HC, and HCT Series of Logic Circuits
EIA-PDP-100	Registered and Standard Mechanical Outlines for Electronic Parts

IPC-SM-786   Recommended Procedures for Handling of Moisture Sensitive Plastic IC Packages

Components, Electromechanical

EIA-506	Dimensional and Functional Characteristics Defining Sockets for Leadless Type A Chip Carriers
EIA-507	Dimensional Characteristics Defining Edge Clips
EIA/IS-47	Contact Termination Standard for Surface Mount Devices

Components, Switches

IECQ-PQC-41	Detail Specification, Dual In-Line Switch, Surface Mountable, Slide Actuated
EIA-448-23	Surface Mountable Switches, Qualification Test
EIA-520EAAAA	Detail Specification for Surface Mountable Dual In-Line Switches of Certified Quality

Components, Tape and Reel

EIA-481-A	Taping of Surface Mount Components for Automatic Handling
EIA-481-1	8mm and 12mm Taping of Surface Mount Components for Automatic Handling
EIA-481-2	16mm and 24mm Embossed Carrier Taping of Surface Mount Components for Automatic Handling
EIA-481-3	32mm, 44mm, and 56mm Embossed Carrier Taping of Surface Mount Components for Automatic Handling

Printed Circuit Boards and Substrates

IPC-FC-250	Performance Specification for Single- and Double-Sided Flexible Printed Boards
IPC-RB-276	Performance Specification for Rigid Printed Boards
IPC-SD-320	Performance Specification for Rigid Single- and Double-Sided Printed Boards
IPC-ML-950	Performance Specification for Multilayer Printed Boards

232   Applied Surface Mount Assembly

MIL-P-50884   Military Specification, Flexible and Rigid Flex, Printed Wiring
MIL-P-55110   Military Specification, General Specifications for Printed Wiring Boards
IPC-L-108   Specifications for Thin Laminate Metal-Clad High-Temperature Multilayer Printed Boards
IPC-L-109   Specifications for Glass Cloth, Resin Preimpregnated (B-Stage) for High-Temperature Multilayer Printed Boards
IPC-L-115   Specification for Plastic Sheet Laminated Metal-Clad for High-Temperature Performance Printed Boards
IPC-CF-148   Resin-Coated Metal for Multilayer Printed Boards
IPC-CF-150   Copper Foil for Printed Wiring Applications
IPC-CL-152   Metallic Foil Specification for Copper/Invar/Copper for Printed Wiring and Other Related Applications
IPC-RF-245   Performance Specification for Rigid Flex Multilayer Printed Boards
IPC-MC-324   Performance Specification for Metal Core Boards
IPC-HM-860   Performance Specification for Hybrid Multilayer

Printed Circuit Board and Substrate Design

IPC-T-50   Terms and Definitions for Electronic Interconnections
IPC-CM-78   Surface Mount and Interconnecting Chip Carrier Guidelines
IPC-H-855   Hybrid Microcircuit Design Guide
IPC-D-249   Design Standard for Flexible Single- and Double-Sided Printed Boards
IPC-D-317   Design Standard for Electronic Packaging, Utilizing High-Speed Techniques
IPC-D-319   Design Standards for Rigid Single- and Double-Sided Printed Boards
IPC-SM-782   Surface Mount Land Patterns
IPC-D-859   Design Standard for Multilayer Hybrid Circuits
IPC-D-949   Design Standard for Rigid Multilayer Printed Boards
IPC-D-275   Design Standard for Rigid Printed Boards and Rigid Printed Board Assemblies
MIL-STD-275   Military Standard Printed Wiring for Electronic Equipment
MIL-STD-2118   Design Standard for Flexible Printed Wiring

EIA-CB-11	Guidelines for the Surface Mounting of Multilayer Ceramic Chip Capacitors
IPC-CM-770	Guidelines for Printed Board Component Mounting
IPC-SM-784	Guidelines for Direct Chip Attachment
SMC-TR-001	An Introduction to Tape Automated Bonding and Fine- Pitch Technology
IPC-SM-780	Electronic Component Packaging and Interconnection with Emphasis on Surface Mounting
IPC-CC-830	Qualification and Performance of Electrical Insulation Compounds for Printed Board Assemblies
IPC-SM-840	Qualification and Performance of Permanent Polymer Coating for Printed Boards

Soldering Materials and Solderability

IPC-SM-817	General Requirements for Surface Mount Adhesives
IPC-SF-818	General Requirements for Electronic Soldering Fluxes
IPC-SP-819	General Requirements for Electronic-Grade Solder Paste
MIL-HM-14256	Flux: Soldering, Liquid, Rosin-Based
EIA/IS-46	Test Procedure for Resistance to Soldering for Surface Mount Components
EIA/IS-49-A	Solderability Test Method for Leads and Terminations
EIA-448-19	Method 19 Test Standard for Electromechanical Components—Environmental Effects of Machine Soldering Using a Vapor-Phase System
IPC-TR-462	Solderability Evaluation of Printed Boards with Protective Coatings Over Long-Term Storage
IPC-TR-464	Accelerated Aging for Solderability Evaluations
IPC-S-804	Solderability Test Method for Printed Wiring Boards
IPC-S-805	Solderability Test for Component Leads and Terminations
IPC-S-815	General Requirements for Soldering Electronic Interconnections
IPC-S-816	Troubleshooting for Surface Mount Soldering
IPC-AJ-820	Assembly and Joining Handbook
MIL-STD-2000A	Standard Requirements for Soldered Electrical and Electronic Assemblies
ANSI/J-STD-001	Requirements for Soldered Electrical and Electronic Assemblies (supersedes IPC-S-815)
ANSI/J-STD-002	Solderability Tests for Component Leads, Termina-

	tions, Lugs, Terminals, and Wires (supersedes IPC-S-805)
ANSI/J-STD-003	Solderability Tests for Printed Circuit Boards (supersedes IPC-S-804)
ANSI/J-STD-004	Requirements for Soldering Fluxes (supersedes IPC-SF-818)
ANIS/J-STD-005	General Requirements and Test Methods for Electronic-Grade Solder Paste (supersedes IPC-SP-819)
ANSI/J-STD-006	General Requirements and Test Methods for Soft Solder Alloys and Fluxed and Nonfluxed Solid Solders for Electronic Soldering Applications

Quality and Reliability

EIA-469-B	Standard Test Method for Destructive Physical Analysis of High Reliability Ceramic Monolithic Capacitors
EIA-510	Standard Test Method for Destructive Physical Analysis of Industrial-Grade Ceramic Monolithic Capacitors
IPC-A-600	Acceptability of Printed Boards
IPC-A-610	Acceptability of Printed Board Assemblies
MIL-STD-883	Methods and Procedures for Microelectronics
IPC-A-38	Fine-Line Round-Robin Test Pattern
IPC-A-48	Surface Mount Artwork
IPC-SC-60	Postsolder Solvent Cleaning Handbook
IPC-AC-62	Postsolder Aqueous Cleaning Handbook
IPC-AI-641	User Guidelines for Automated Solder Joint Inspection Systems
IPC-AI-642	User Guidelines for Automated Inspection of Artwork and Innerlayers
IPC-AI-643	User Guidelines for Automatic Optical Inspection of Populated Packaging and Interconnection Assemblies
IPC-SM-785	Guidelines for Accelerated Surface Mount Attachment Reliability Testing
EIA-JEDEC	Method B 105-A, Lead Integrity—Plastic Leaded Chip Carrier Packages
EIA-JEDEC	Method B 102, Surface Mount Solderability Test
EIA-JEDEC	Method B 108, Coplanarity
IPC-TM-650	Test Methods Manual

Equipment Requirements

SMEMA 1.1   Mechanical Equipment Interface Standard
SMEMA 2.0   Software/Communications Interface Standard
SMEMA 3.1   Fiducial Mark Standard

Index

Activator, definition of, 67
Adhesives
  acrylics, 68
  curing, 68, 69
  epoxies, 68
  glass transition temperature, 68
  viscosity, 69
Adhesive curing, 162
Adhesive printing, 98
Alloy 42 leads, 80, 203
Alloy, definition of, 67
Aperture design, 99
Aperture, definition of, 83
Aqueous cleaning, 191
  batch equipment, 192
  definition of, 185
  in-line equipment, 193
Aqueous cleaning,

Barrel, definition of, 47
Batch flow manufacturing, 218
  definition of, 213
Bi-level stencil, 102
Bidirectional wave form, definition of, 165
BQFP, 31, 32, 33
BQFP land pattern, 32
Breakaway tabs, definition of, 59
Brookfield viscometer, 76, 69
Bumpered quad flat pack, definition of, 9

CCD camera, definition of, 115
Centi-poise, 69, 76
Ceramic leaded chip carrier, 39
  definition of 9
Chip component, definition of, 9
Chip resistors, 11, 12, 13
Chip resistor land pattern, 14
Chip capacitor, 13, 14, 15
  cross section, 15
  land pattern, 18
  process flow, 16, 17
Class A, definition of, 1

Class B, definition of, 1
Class C, definition of, 1
Cleanliness testing, 197
  electromigration testing, 199
  solvent extraction testing, 198
  surface insulation resistance testing, 198
  visual inspection, 198
Coefficient of thermal expansion, 49, 50, 51
  definition of, 9
Component feeders, 128
  cut and form, 131
  gravity, 129
  tape and reel, 130
  tray, 131
  vibratory, 130
Component hole, definition of, 47
Component packaging, 42, 43, 44, 45, 61
Component spacing, 61, 62
Component taping, 131
  equipment, 132
  materials, 131
Conduction, definition of, 135
Contact angle, 79
Contact printing, 93
Contact soldering, definition of, 201
Contaminants, 186
Continuous flow manufacturing, 219
Continuous-flow manufacturing, definition of, 213
Convection reflow, definition of, 135
Convection, definition of, 135
Coplanarity, definition of, 47
Crystals, 42
Copper leads, 80
Cylindrical component, definition of, 9

Defect, definition of, 201
Delamination, definition of, 47
Dewetting, 79, 80
DFA, 60
Dielectric, definition of, 9

237

238 Index

Diffusion, definition of, 67
DIP switches, 40
DIP switch land pattern, 41
Dispensing, 106
  attributes, 84
  concerns, 84
  defects, 110
  methods, 107
DPAK, 24
DPAK land pattern, 26
Dross, definition of, 165
Dry-film, definition of, 47
Dual in-line package, definition of, 1
Dual wave system, definition of, 165
Ductility, definition of, 47
Dwell time, definition of, 165

Economic lot size, 214
Electroless plating, 52
  definition of, 47
Epoxy smear, definition of, 48
EIA, 229
Encoder, definition of, 115
Entry angle, definition of, 165
Eutectic alloy, definition of, 67

Fiducial, 64, 65, 66, 112, 128
  definition of, 59
Fine pitch, definition of, 10
Fixed squeegee, definition of, 83
Flexible stencil, definition of, 83
Floating squeegee, definition of, 83
Flow lines, 220
Flux, 69
  activation, 72
  classification, 69
  copper mirror test, 70
  corrosion test, 70
  groups, 71
  in solid paste, 72
  low solids, 72
  process window, 70
  rosin based, 71
  silver chromate test, 70
  surface insulation resistance testing, 70
  synthetic activated, 71
  water soluble, 71
Flux module, definition of, 165
Flux, definition of, 67
Forced air convection reflow, 152
  attributes, 155
  concerns, 156
  conductive convection, 154
  infrared convection, 153
Freeboard, definition of, 185

Glass transition temperature, definition of, 48
Gull wing lead, definition of, 10

Hole size, 63
Hot air leveling, definition of, 48
Hot gas soldering, definition of, 201

IEC, 229
Infrared reflow, 146
  definition of, 135
  lamp IR systems, 147
Infrared soldering, definition of, 201
Integrated circuit, definition of, 1
Intermetallic layer, 79, 202
  definition of, 67
Ionic contamination, definition of, 185
IPC, 229

JEDEC, 229
J-lead, definition of, 10

Kanban, 219
  definition of, 213
Kovar leads, 80, 202

Laminar wave form, definition of, 165
Laminate, definition of, 48
Lamp IR attributes, 148
Lamp IR concerns, 148
  panel IR attributes, 150
  panel IR concerns, 152
  panel IR systems, 149
Land, definition of, 1
Land pattern, definition of, 59
Leaching, 80, 202
Leaching, definition of, 67
Lead pitch, definition of, 10
Leadless ceramic chip carrier, definition of, 10
Liquid photoimageable solder mask, definition of, 48
Liquidus, definition of, 67
Lot size, definition of, 213
Lot, definition of, 213

Maintenance, 226
Malcom viscometer, 76
Material handling equipment, 222
MELF land pattern, 21
Metal electrode leadless face, definition of, 10
MIL, 229
Minimum lot size, 214
Molded capacitor land pattern, 20
Montreal protocol, 186
Multilayer chip capacitor, definition of, 10

Nickel barrier, 80, 203
Noneutectic alloy, definition of, 67
Nonwetting, 79, 80

Off-contact printing, 93
  definition of, 83

Index 239

On-contact printing, definition of, 83
Opacity, definition of, 135
Ozone depletion potential, definition of, 185

Pad, definition of, 1
Pad size, 63
PCA, definition of, 1
PCB, definition of, 1
PCB, panels, 55, 57
Placement
  actual rate, definition of, 115
  classification, 117
  component centering, 125
  equipment design, 117
  external, definition of, 115
  heads, 124
  internal, definitions of, 115
  maximum rate, definition of, 115
  mechanical, 126
  repeatability, definition of, 115
  vision, 127
Plastic leaded chip carrier, definition of, 10
Plated through hole, definition of, 48
PLCC, 28, 29, 30
  land pattern, 30
  sockets, 40, 41
Poise, 69, 76
  definition of, 67
Polar contamination, definition of, 185
Position systems, 125
Preheat module, definition of, 165
Primary side, definition of, 48
Print head assembly, 94
Print modes, 92
Printing, 84
  attributes, 84
  concerns, 84
  defects, 104
  off-contact, 84
  on-contact, 84
  parameters, 102
  PCB design, 103
Profile, definition of, 135, 165
Profiling issues, 137
  components, 140
  flux, 139
  printed circuit boards, 137
  solder, 140
  time and temperature, 137

QFP, 31, 32, 33
  land pattern, 34
Quad pack, definition of, 10

Radiation, definition of, 135,
Rectangular component, definition of, 10
Reflectivity, definition of, 135

Reflow
  control systems, 156
  controlled atmospheres, 159
  profiling, 159
  transfer systems, 157
Reflow temperature, definition of, 135
Rework
  component concerns, 202
  contact soldering, 203, 205, 206
  flux and solder, 210
  heat stress, 210
  hot gas soldering, 206, 208
  infrared soldering, 208, 209
  jumper wires, 210
  thermal shock, 202
  workmanship standards, 211

Saponifier, definition of, 185
Semiaqueous cleaning, definition of, 185
Screen, definition of, 83
Screens, 85
Secondary side, definition of, 48
Semiaqueous cleaning, 194
  batch equipment, 194
  in-line equipment, 196
  solvents, 194
Servo motor, definition of, 115
Setup, 217
Shrink small integrated circuit, definition of, 10
Silk screen, 55
Small outline J-leaded, definition of, 10
Small outline large integrated circuit, definition of, 10
Small outline transistor, definition of, 10
Small outline integrated circuit, definition of, 10
SMC, definition of, 1
SMT classifications, 5, 6
Snap off, definition of, 83
SOIC, 24, 25, 26
SOIC land pattern, 27
SOJ, 26, 27
Solder, 72
  binary systems, 73
  creep, 73
  diffusion, 73
  dross, 74
  eutectic, 73
  leaching, 74
  liquidus, 73
  noneutectic, 73
  reflow soldering alloys, 74
  solidus, 73
  ternary systems, 73
  unary systems, 73
  wave soldering alloys, 73
Solder mask, 53, 54
Solder module, definition of, 165

Solder paste, 75
  characteristics, 75
  percent metal, 77
  powder shape, 77
  powder size, 77
  slump, 76
  tack time, 76
  viscosity, 76
Solder paste printing, 100
Solder paste, definition of, 68
Solder, definition of, 68
Soldering, definition of, 68
Solidus, definition of, 68
Solvent cleaning, 188
  batch equipment, 190
  definition of, 185
  in-line equipment, 190
  solvents, 188
SOT23, 21
  land pattern, 23
SOT89, 22
  land pattern, 24
SOT143, 23
  land pattern, 25
Squeegee blade, 89
  definition of, 83
  metal, 91
  plastic, 90
SSOP land pattern, 38
Statistical process control, definition of, 1
Stencil, definition of, 83
Stencils, 87
  aspect ratio, 88
  etch factor, 88
Stepper motor, definition of, 115
Surface mount adhesive, definition of, 68
Surfactants, definition of, 185

Tantalum capacitor cross section, 19
Tapepak, definition of, 10
Tensile strength, definition of, 48
Termination, definition of, 11
Tg, 49, 50, 51, 138
Thermal shock, definition of, 201
Thermocouple, 160
  attachment, 161
  definition of, 135
  junction, 161
  placement, 162
Thermoplastic, definition of, 68

Thermoset, definition of, 68
Thin small outline package, definition of, 11
Tombstoning, definition of, 135
Tooling holes, definition of, 48
Translucency, definition of, 135
Transparency, definition of, 136
TSOP, 35
TSOP Type I land pattern, 36
TSOP Type II land pattern, 36

Vapor phase reflow, definition of, 136
Vapor phase reflow, 141
  attributes, 145
  batch systems, 142
  concerns, 145
  continuous systems, 143
  fluid, 141
  preheat, 145
Vehicle, definition of, 68
Via hole, definition of, 48
Vibrating wave, definition of, 165
Viscosity, definition of, 68
Vision alignment, 112
Volatile organic compound, definition of, 165

Waste water management, 197
Wave soldering
  bidirectional wave form, 177
  component layout, 170
  components, 169
  definition of, 48
  dross control, 180
  dual wave, 178
  flux, 169
  flux application, 172
  flux density control, 174
  inert atmospheres, 180
  laminar wave form, 177
  PCB handling, 181
  preheat, 174
  printed circuit boards, 168
  profiles, 166
  solder application, 176
  vibrating wave, 179
Wetting, 78
Work cell, definition of, 213
Work cells, 220

Zone, definition of, 136

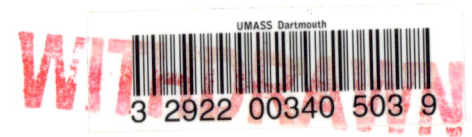